TAR SANDS
SHOWDOWN

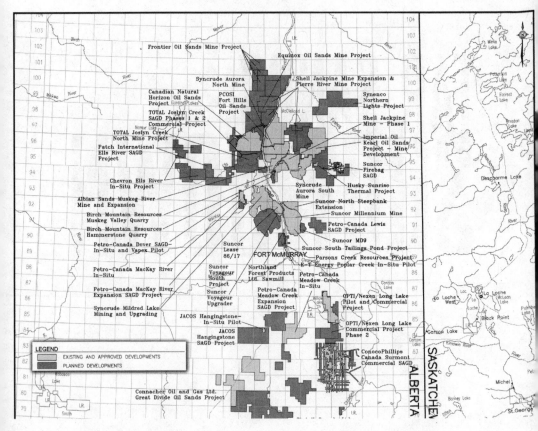

Major oil company land leases and plant sites in the Athabasca region of the tar sands. Oil sands base map produced by Golder Associates Ltd., August 2008.

CANADA AND THE NEW POLITICS OF OIL
IN AN AGE OF CLIMATE CHANGE

TAR SANDS
SHOWDOWN
TONY CLARKE

JAMES LORIMER & COMPANY LTD., PUBLISHERS

James Lorimer & Company Ltd., Publishers acknowledge the support of the Ontario Arts Council. We acknowledge the support of the Government of Canada through the Book Publishing Industry Development Program (BPIDP) for our publishing activities. We acknowledge the support of the Canada Council for the Arts for our publishing program. We acknowledge the support of the Government of Ontario through the Ontario Media Development Corporation's Ontario Book Initiative.

Library and Archives Canada Cataloguing in Publication

Clarke, Tony
Tar sands showdown : Canada and the politics of oil in
an age of climate change / Tony Clarke.

Includes bibliographical references and index.
ISBN 978-1-55277-018-4

1. Oil sands industry — Environmental aspects — Canada.
2. Oil sands industry — Economic aspects — Canada. 3. Climatic
changes — Environmental aspects — Canada. 4. Oil sands industry — Environmental aspects
— Alberta. 5. Oil sands industry — Economic
aspects — Alberta. 6. Oil sands industry — Social aspects — Alberta.
I. Title.
HD9574.C23A54 2008   333.8'2320971   C2008-905707-4

Cover design by Meghan Collins

Tar Sands Showdown is printed on Silva Enviro 100% recycled paper.

James Lorimer & Company Ltd., Publishers
317 Adelaide Street West, Suite 1002
Toronto, Ontario
M5V 1P9
www.lorimer.ca

Printed and bound in Canada

CONTENTS

Preface 7

Acknowledgements 12

CHAPTER 1: Crude Awakening 14

CHAPTER 2: A Triple Crisis 48

CHAPTER 3: Energy Superpower 79

CHAPTER 4: Fuelling America 114

CHAPTER 5: Ecological Nightmare 148

CHAPTER 6: Social Upheaval 183

CHAPTER 7: Resistance Movement 217

CHAPTER 8: Dream Change 248

TAR SANDS PROBE: A Brief Study Guide 282

Notes 289

Selected Bibliography 306

Index 307

To Hugh McCullum, whose lifelong dedication to the struggle of indigenous peoples, both at home and in the world, has been an inspiration.

Preface

When I reflect on it, my interest in the Alberta tar sands probably dates back more than thirty years. It was during the mid 1970s that I first read Larry Pratt's seminal work, *The Tar Sands: Syncrude and the Politics of Oil.* As an activist and later as an author, I was intrigued by the challenges of the emerging tar sands puzzle. Although I was stimulated by Pratt's analysis, I was rather heavily preoccupied with other social struggles and had no intention of taking on another issue, let alone something as complex as the tar sands. Nor did I have any intention of writing about the tar sands, let alone an entire book. Yet my interest in the subject grew when I visited the tar sands region of Alberta a few times and could see first-hand the developments taking place.

During the past three decades, I made three trips from Edmonton to Fort McMurray for onsite visits of the tar sands, including the Syncrude and Suncor plants. With each trip, I felt the magnitude of these colossal enterprises and their ecological footprints. Each visit, I was increasingly awestruck by the sheer scope and size of the sophisticated machinery and technology being used to mine or siphon the bitumen from the earth's crust to produce the oil. But, more so, I was shocked and stunned by the degree of environmental devastation, ranging from the strip mining of the boreal forests and the massive use of water sources to the creation of huge toxic tailings ponds the size of lakes. And, always, I would take a moment to ponder the fact that this tar sands mega-machine primarily existed in order to satisfy an increasingly oil-thirsty America.

With each trip, there were particular moments of reflection. During the first trip in 1976, I was working with Project North, a coalition of the country's major national churches, to build public support in southern Canada for the struggle of the Dene Nation against the construction of the Mackenzie Valley Pipeline. The pipeline corridor was designed to bring natural gas from the Beaufort Sea across the indigenous lands of the Dene in the Northwest Territories to markets in southern Canada (including the tar sands industry) and the United States. What I saw happening in the tar sands back then, though they were still in early stages of development, only strengthened my resolve to support the Dene campaign against the pipeline. My second onsite visit to the tar sands occurred in 1997, several years after I chaired the Action Canada Network, the nation-wide coalition that waged the historic fight against Canada's free trade deal with the US and the North American Free Trade Agreement (NAFTA). Since Canada had effectively surrendered its sovereignty over the country's energy resources through the new free trade regime, there were clear and ominous signs then that Alberta and the tar sands industry would soon be America's leading energy satellite as long as NAFTA remained intact.

However, by the time I made my third trip to the Alberta tar sands a decade later, the world had undergone some profound and significant changes, thereby adding new dimensions and challenges. Under the Bush administration in Washington, energy had become a national priority and, following the attacks of September 11, 2001, one of the top issues, if not the top issue, of national security for the US. The US-led invasion of Iraq, which arguably had more to do with securing greater control over Middle East oil deposits than fighting the war on terror, highlighted the need to focus more attention on how the US military is geared to policing the world in order to control oil sources and supply routes to serve its energy interests. At the same time, Canada via the Alberta tar sands surpassed Saudi Arabia as America's number-one foreign supplier of crude, thereby situating this country in a new geopolitical context as far as the politics of oil is concerned. Suddenly, Canada had become a global player, with the tar sands being the second-largest deposit of recoverable crude oil in the world.

Yet, the most dramatic transformation came in terms of a growing recognition of the ecological limits of the planet. Climate change had risen to the top of the global agenda. Rapid global warming, due mainly to the burning of fossil fuels that releases greenhouse gases into the atmosphere, thereby trapping heat on the planet, is having a profound impact on both the petroleum industry and national politics. Not only is the petroleum industry one of the main sources of fossil fuels, but tar sands oil production generates more than three times the amount of greenhouse gas emissions as conventional oil production. What's more, there is also an increasing global awareness that many of the natural resources of the planet, such as fresh water, are rapidly depleting or becoming contaminated. Since the production of oil from the tar sands uses and pollutes a great deal of fresh water to extract the bitumen from the earth's crust, the industry is now becoming a prime example of the abuse of water sources.

All of this is taking place at a time when the world is on the verge of running out of sufficient supplies of oil to meet growing global demand. Many geologists believe that the period of abundant and cheap supplies of oil is over. The production of conventional sources of oil has either peaked, or is about to peak, while remaining oil reserves are the hard-to-get-at sources like the tar sands, which are much more expensive, economically and environmentally, to develop. At first glance, this may seem to be good news for Canada and the Alberta tar sands. But, even all-out production of the tar sands will only make up for a fraction of the oil shortages forecasted. For Canada, as well as other industrialized countries that have built their economies on the availability and abundance of cheap supplies of oil, the era of peak oil will come as a shock. And, since Canada has virtually surrendered control over our energy resources through the free trade regimes, Canadians would do well to think again about how vulnerable we could be in this new era of peak oil.

In short, the development of the tar sands poses a whole new set of critical challenges for Canada at this moment in history. The tar sands are, in many respects, a microcosm of the economic forces and ecological challenges facing the future of our industrialized society and way of life. How Canada manages the development of the tar sands will determine, in large measure, our destiny as a nation in the twenty-first century. We can continue our headlong rush to develop the tar sands for massive exports of dirty crude oil to the US, thereby establishing this country as, in the words of the current prime minister, the world's next energy superpower, but leaving a legacy of ecological and social damage in our wake. Or, we can choose to embark on an alternative path of developing a made-in-Canada energy and environmental strategy that secures public control over the development of the tar sands with a view to phasing out production and making a deliberate transition to a new energy future based on renewable resources rather than fossil fuels.

This is why the time has come for a showdown over the tar sands. The development of the tar sands is no longer an Alberta issue alone but very much a national and even a binational one. After all, the Alberta tar sands are more than just another mega-energy project. As our leading high-priced resource commodity, crude oil has become the cornerstone of the Canadian economy, exercising a decisive influence on currency markets, manufacturing exports and the country's overall trade balance. Playing this central role in the Canadian economy also means that the tar sands enterprise will, whether we like it or not, shape much of our direction and destiny as a nation for the foreseeable future. If the tar sands are interwoven with the destiny of the country, then we all as Canadians have a stake in the outcome. In effect, the rest of Canada must share responsibility with Alberta for the development of the tar sands and its consequences.

The prime purpose of this book is to help stimulate a nationwide public debate about the tar sands and the critical issues at stake concerning Canada's future. In doing so, the objective is to cast the development of the tar sands in the context of a broader set of forces and challenges, thereby providing a multifaceted lens for observing, judging and acting. The development of the tar sands poses fundamental challenges for Canada. It's time to put aside the platitudes and confront the hard and difficult issues head-on. Continuous denial of our responsibilities as a nation concerning the tar sands and their implications for Canada's role in this century is no longer an option. Now we must put aside the illusion of becoming the "world's next energy superpower" and start dealing with the new politics of oil in an age of climate change. Let us seize this moment by confronting the tar sands challenge and developing an authentic energy and environment strategy for a more sustainable future.

August 2008

Acknowledgements

As always, there have been many people who have contributed directly or indirectly to the project of writing this book. Several are identified in the manuscript, while others have been thanked personally or referenced in the footnotes.

In particular, I want to thank Jim Lorimer and Bruce Campbell for recognizing and believing in the importance of co-publishing a book on the tar sands at this time. Their support for this project has been unwavering. Much appreciation also goes to Lynn Schellenberg at James Lorimer & Co. for coordinating the publishing of the book and special thanks to Pamela Martin for her expert guidance in the editing stages of the manuscript.

I am very grateful to several people who studiously read various chapters and gave me specific feedback and suggestions. They include Hugh McCullum and John Dillon, with whom I have worked closely in the past on energy issues and related social struggles, as well as two of my co-workers at the Polaris Institute,

Richard Girard, our research coordinator, and Jessica Kalman, our campaigner on tar sands issues (www.tarsandswatch.org). Patricia Flores, our office manager at Polaris, was also very helpful when my computer crashed in the midst of editing the manuscript.

Other important insights and contributions came in the form of collaborative work and conversations with people from a variety of groups and organizations active on tar sands issues, including Gordon Laxer, Diana Gibson, Dan Woynillowicz, Lindsay Telfer, Simon Dyer, Stephen Hazel, Matt Price, Bill Moore-Kilgannon, Gil McGowan, Ricardo Acuna, Chris Burn, Jim Stanford, Mel Watkins and Cy Gonick.

I am also grateful to friends and colleagues in the US and elsewhere who, in varying ways, encouraged me to write this book and contributed insights, including Osprey Orielle Lake, Tom Goldtooth, Jerry Mander, Michael Klare, Jack Santa-Barbara, Mary Anne Hitt, Susan Casey-Lefkowitz, Liz Barratt-Brown, Steve Kressman, Randy Hayes, Richard Heinberg and David Korten.

Finally, a special word of thanks to my family — my wife, Carol, who continues somehow to put up with me through writing projects like this, and our two children, Tanya and Chris, who, along with their partners, are part of a generation that is increasingly challenged by many of the issues discussed in this book.

CHAPTER 1

Crude Awakening

The end of cheap whale oil marked the beginning of the age of black gold. From its humble beginning as fuel for lamps, oil became *the* energy source of the twentieth century. Although coal produced more total energy over the century, and generated a huge share of our heat and power, it was oil that fuelled the new modes of transportation — the automobile, railways, ships and airlines — that shaped much of the century. These developments, in turn, led to a myriad of new oil-based products, from plastics to synthetic rubber, which further reinforced the demand for oil and its increasing value as a form of energy. In short, oil became the lifeblood of industrial civilization and remains the black gold of the twenty-first century.

Although US historians generally claim that it was Edwin L. Drake who drilled the first oil well, on August 27, 1859, in the town of Titusville, Pennsylvania, the beginning of commercial oil production is essentially a Canadian story. A year earlier, in the summer of 1858, a carriage maker by the name of James Miller Williams drilled an oil well fifteen metres down into the oil gum beds near Black Creek in Lambton County, southwestern Ontario. Williams's well marked the start of the first black-gold rush, which lasted three decades. The area surrounding Black Creek (later renamed Oil Springs) was stripped of its wetlands and forests to make way for some 1,600 oil-drilling rigs. Dozens of oil companies (including Imperial Oil) and refineries sprang up during this period, thereby giving birth to Canada's oil and gas industry.[1] Over the next 120 years, as the demand for oil grew, other major oil deposits were discovered and tapped, including at Leduc, near Edmonton, and at Weyburn in southern Saskatchewan in 1947, and finally, towards the end of the discovery of major deposits of conventional oil on dry land in Canada, offshore discoveries were made at Sable Island in Nova Scotia in 1970 and at Hibernia in Newfoundland in 1979.

Now, 150 years later, Canada is at the centre of another major black-gold rush, this time in the Athabasca tar sands of Alberta. However, this is not what has become known over the past century and a half as conventional oil production, i.e., pumping crude oil out of the ground. Much of the earth's oil that lies close to the surface has already been used up and discoveries of new oil that is cheaply accessible have become fewer and fewer. Instead, the oil industry is now having to turn its attention more and more to the harder-to-get-at oil deposits that lie deeper within the earth's crust, in deposits that are more difficult to extract and forms that require more processing, such as in the Alberta tar sands, or under the sea. This unconventional oil production requires new technologies and is more expensive to produce. But the rapid rise in world oil prices in recent years (more than doubling between June 2007 and June 2008) has made the development of the Alberta tar sands a very lucrative business.

This photo shows the boreal forest region alongside tar sands production areas. Courtesy David Dodge, Canadian Parks and Wilderness Society.

Black Gold

Today's black-gold rush has been many years in the making. The hub of the Alberta tar sands, at the point where the Athabasca and the Clearwater Rivers meet in northern Alberta, has been the home of First Nations since time immemorial. In 1778 American explorer Peter Pond travelled northward down the Athabasca to become the first "white man" to see the "springs of bitumen which flow along the ground." For the indigenous peoples, however, this bitumen or "black glue" was well known. They boiled it to extract fuel for heating and to provide caulking to seal the seams of their birchbark canoes. It was also used for medicinal purposes. Indeed, this oil-impregnated earth appears to have been an integral part of the First Nations' way of life in the Athabasca boreal.

Ten years after Peter Pond paddled the Athabasca, his better-known successor, Alexander Mackenzie, followed the same route in his search for a Northwest Passage. Paddling down the Athabasca, he came across what the Dene call the Dehcho River (which, translated, means the great river), later named the Mackenzie River by the English after his expedition. Marvelling at the strange, thick layers of tar sands along the banks of the Athabasca, Mackenzie wrote in his diary: "At about twenty-four miles from the Fork are some bituminous fountains, into which a pole of twenty feet long may be inserted without the least resistance. In its heated state, it [the bitumen] emits a smell like that of sea coal. The banks of the river, which are there very elevated, discover veins of the same bituminous quality."[2]

As Pond and Mackenzie traversed the Athabasca lands of the First Nations, they stumbled upon what is now considered to be one of the richest, if not the richest, reservoirs of untapped oil to be found anywhere in the world. The tar sands, or, as the industry calls them, the oil sands, are composed of bitumen, a very heavy oil with a high sulphur content, surrounded by a mixture of sand, clay, minerals and water. Each grain of sand is surrounded by a water envelope that keeps it separated from the oil. Without this water envelope, the oil and sand could not be separated. A clump of tar sand actually looks and feels like asphalt. While there are deposits of tar sands close to the surface in some areas north of Fort McMurray, most of them are buried deep in the earth, up to 610 metres below the surface. Some of the largest deposits of tar sands are 50 metres thick, or more.

The geological origins of the tar sands are a bit of a mystery. How these oil-impregnated sands were formed over the course of millions of years is still being debated among geologists. Some say that originally the bitumen was a lighter form of oil that gradually

migrated from other strata to its present location where it bitu-
minized in the sands basin to become a tar-like substance. Others
maintain that the bitumen in the tar sands is really a kind of
proto-petroleum unique to the sands strata in which it is found
and in the early stages of transforming into the lighter petroleum
with which we are more familiar.[3] Whatever its geological origins,
however, the bitumen located in the massive Athabasca hydrocarbon deposit does resemble the heavy conventional oil found
deeply buried elsewhere in the Alberta oil patch. Similar deposits
of tar sands are also found in Saskatchewan, where it is being
developed more quietly, and in Venezuela.

As hydrocarbons go, bitumen is not particularly attractive. In
its raw state, it is too thick to transport by pipeline. It also has a
lower rate of hydrogen to carbon than conventional oils. And, it
possesses a greater sulphur content. But through processing and
upgrading, bitumen can be broken down and coked, depleting
sulphur and adding hydrogen content, to yield a high-quality
blend of naphtha and light and heavy gas oils.[4] Although the oil
industry calls this blend "synthetic crude oil," it is not identical to
conventional crude oil. As various observers have noted, the
blend is actually of higher quality than ordinary crude and has
more potential uses, especially in the production of petrochemical products. Yet it is refined, marketed and priced along with
conventional oil as part of the system of petroleum production in
North America.

Alberta's massive tar sands are located northeast of Edmonton
(in three fields), about 600 kilometres straight north of the
Canada–US border. Estimates put the total amount of potential
crude oil in the tar sands at a whopping 2.5 trillion barrels. However, only a portion is recoverable. Given current technologies,
official global estimates say there are between 175 and 200 billion
barrels of recoverable oil there, a reserve second only to that of
Saudi Arabia. However, working from slightly different assump-

tions and a different notion of "recoverability," the Canadian Association of Petroleum Producers estimates that there are some 315 billion barrels of recoverable crude oil in the tar sands, a greater reserve than that of Saudi Arabia. In addition, there are an estimated 1.1 billion barrels of recoverable heavy oil reserves in Saskatchewan, which are considered by geologists to be part of the tar sands.

The tar sands are located in three separate fields, the largest of which is the Athabasca, followed by Cold Lake and the Peace River (Figure 1.1). Together, the three main fields cover an area totalling 141,000 square kilometres, which is larger than the area of the Maritime provinces combined. In order to develop the tar sands, however, a major riddle had to be solved: how could bitumen be made, in Pond's words, to "flow like molasses in the hot sun." At first, it was thought that the oil could be drilled out of the black sands of Alberta. After all, if oil was oozing to the surface as it did in the gum beds of Lambton County or Pennsylvania, then there must be deposits of oil underneath. In the 1890s, Ottawa sent two separate drilling expeditions to the tar sands of Alberta, but no such deposits were found. After these two failed expeditions, Ottawa sold leases to private prospectors in hopes that they would find a solution.[5] Lured by stories of the black-gold rush, promoters and prospectors alike, often colourful figures, trekked into the Athabasca forests to try to pump out the bitumen or melt it with fire to separate it from the sands. Soon it became clear that the oil-laden tar sands were different than conventional oil deposits and new methods had to be found for extracting the oil. Each clump of clay and sand contains roughly 10 to 12 percent bitumen. The great unresolved riddle was how to separate the bitumen from the sand and clay.

When the prospectors and speculators failed to find a solution, Ottawa's Mines Branch focused its attention on the great tar sands riddle. In 1913, Sidney Ells, a vigorous and determined

Figure 1.1. Oil Sands Fields in Alberta.
Source: Rick Schneider, Canadian Parks and Wilderness Society.

federal mining engineer and surveyor, led a scientific expedition into the Athabasca tar sands under some of the harshest conditions imaginable. Flanked by a crew made up mainly of Athapaskans, including Dene, from the region, Ells laboriously trekked through the brush and along the shores of the Athabasca, gathering samples and surveying the bitumen deposits. Ten tonnes of samples of the bituminous sand were loaded onto scows and shipped up the river to Fort McMurray, and from there on to Ottawa by train. Back in Ottawa, Ells conducted tests on the samples and eventually came to the conclusion that heat is the solution to the tar sands riddle. By applying hot water and steam, the bitumen can be separated from the sand.

In 1914, Ells compiled his research in a 110-page report entitled *Notes on Certain Aspects of Proposed Commercial Development of the Deposits of Bituminous Sands in the Province of Alberta.*[6] His report mapped out where the richest veins of oil-soaked sand could be found, gave the molecular and chemical content of the tar sands, and described the method he found for separating the bitumen from the sands. Ells's report also included photographs he had taken of the Athabasca region, and his recommendation that the best way to get at the most readily available bitumen deposits was through mining. In many respects, Ells's report served as a road map for tar sands companies, such as Syncrude and Suncor, that later mined the rich veins of bitumen he had documented.

Around the time that Ells completed his report World War I had begun: he went overseas with the Canadian Armed Forces, not to return until 1919. Meanwhile, his work was picked up by Karl Clark, a chemist with the federal Mines Branch. Using the samples Ells had brought back from the Athabasca, Clark began conducting his own experiments, attempting to extract the bitumen from the sands using chemical solvents. When this failed, Clark reviewed Ells's report and quickly adopted his ideas of using heat and steam for extraction. Using Ells's research as a springboard, Clark designed a washing machine to separate the bitumen from the sand and clay, thereby creating what he called "a buoyant oil froth" that could be skimmed from the surface. Using government grants from both Ottawa and Alberta, between 1920 and 1950 he built three pilot plants using his technology to extract bitumen from the tar sands.

The research carried out by Ells, Clark and their associates laid the foundations for the tar sands industry. Without their scientific and technological breakthroughs the great tar sands riddle would not have been solved. What's more, as Larry Pratt observes, this was "publicly financed research and technical know how" developed through government departments in both Ottawa and Edmonton.[7] In other words, it was the people of Canada and Alberta who, through their governments, invested in the development of this technology. But this is a history that is not without controversy. According to William Marsden, Clark's use of Ells's research on heat and steam extraction was a case of "academic pilfering." While Ells was overseas during the war, Clark "purloined" Ells's findings, then left Ottawa to take a job at the University of Alberta to further develop the technology for extracting oil from the tar sands.[8]

Because bitumen in its natural state cannot be pumped out of the ground the way conventional oil can, it must be made to flow somehow, or the sands must be mined, and the bitumen extracted in a subsequent operation. As a result, two methods were developed.[9] The first was the open-pit or strip-mine method to be used where the tar sands lie close to the surface, i.e., deposits lying up to 60 to 70 metres below the boreal muskeg. But, since only a small percentage (less than 5) of the tar sands lie so close to the surface, other technologies were needed. The method devised to get at the deeper bitumen deposits is called in situ.

The strip-mining method involves three stages. First, the trees and muskeg ("overburden") are stripped away along with the top layers of the earth's crust, laying bare the espresso-coloured sands underneath. Second, the hot water and steam extraction procedure is used to separate the bitumen from the clay, sands and other minerals. Third, the bitumen is diluted, upgraded into synthetic crude, and transported by pipeline to refineries, mainly in the US. Each of these stages required the development of new equipment and technologies. For stage one, gigantic earth removal equipment was needed to shovel and transport massive volumes of overburden, such as the present-day 400-ton Caterpillar 797 whose tires alone are 3.42 metres high. For stage two, plants modelled after the pilot plants designed by Clark and his associates were built to extract the bitumen. Most of the strip-mining projects are concentrated along one piece of the Athabasca deposit located due north of Fort McMurray along both sides of the Athabasca River.

The in-situ method uses completely different technologies to get at the bitumen that lies at deeper levels. Given that between 93 and 95 percent of the tar sands are located at depths of up to 600 metres below the earth's surface, technologies were required to extract the bitumen from the sands in situ, literally "in place." The process involves heating the bitumen and then pumping it out of the ground. Current technologies include the pumping of steam down one drill pipe to separate the bitumen from the sand and clay deep to melt it so that it can be pumped to the surface through a second pipe. Over the past forty years or so, the technologies used for in-situ production have been tested and improved in the expectation that this method will be used much more extensively in the future. But, as we shall see in Chapters 5 and 6, it uses enormous amounts of both natural gas and fresh water, and as a result it has become highly controversial from an environmental standpoint, even more controversial than the strip-mining of the tar sands.

While strip mining and in-situ extraction have emerged as the two main methods used in the development of the tar sands, they are not the only ones that have been proposed. The most startling idea was the brainchild of Manley Natland, an American geologist, who in 1957 proposed that the power of an underground nuclear blast be used to get the bitumen flowing.[10] Natland secured the support of Richfield Oil, a US company that became one of the principal players in the early stages of the Syncrude consortium. At the time, the US Atomic Energy Commission's (AEC) Project Plowshare was exploring the possibility of using nuclear bombs to blast open the earth to create canals, lakes and harbours as well as to facilitate the mining of deeper subsurface ore bodies. After several tests were conducted during this period using underground nuclear explosives, Richfield executives met with government officials in Edmonton and Ottawa in 1958 and gained the support of then Premier Ernest Manning for the project to nuke the tar sands. Soon after, a federal–provincial task force was put together to work out the implementation of the project.

At the same time, Richfield Oil executives were meeting with officials in Washington, including Project Plowshare founder Edward Teller, who had also worked on the infamous Manhattan Project. Teller was convinced that a nuclear explosion in the tar sands would generate sufficient heat to, as he says in his report, "transform the material into a liquid state that can be pumped to the surface."[11] This would be a cheap method for producing enough oil from the tar sands to supply the US's energy needs for centuries to come. By March 1960, the US Congress's Atomic Energy Committee had given its stamp of approval to the project and the US government agreed to supply the nuclear bomb required for the underground blast. Subsequent events, however, put the brakes on the project. The AEC's Gnome Project test explosions in New Mexico had exposed the serious risks of spreading radioactivity. As well, new conventional oil fields were discovered at Prudhoe Bay in Alaska, a source of an abundance of cheap oil within US borders. In addition, the Diefenbaker government in Ottawa banned nuclear testing on Canadian soil. In his 1963 report to the AEC, Teller concluded that the idea of using nuclear devices in the tar sands was premature.[12]

In retrospect, it is difficult to believe that the powers in Washington, Ottawa and Edmonton came so close to pushing the nuclear button on the Athabasca tar sands. Almost half a century later, the technologies developed for strip mining and in-situ production, not nuclear explosives, are leading the way in exploiting the vast potential of the tar sands. But, as we shall see in subsequent chapters, the US demand for Canadian oil has grown exponentially in recent years. While the idea of the use of nuclear explosive devices to make, in the words of Peter Pond, "springs of bitumen ... flow along the ground," seems far-fetched, it is worth noting that the US Secretary of the Navy, representing the Government of the United States, has held the patent for the nuclear technology designed for this purpose since September 1973.[13]

Land Grabs

The prospect of finding and extracting vast amounts of black gold in the Athabasca tar sands set in motion what became the great land grab game. The issue of who owns and who controls the lands in which the tar sands are located underlies a political tug-of-war that has gone on for more than a century. Regardless of the technological breakthroughs facilitating the extraction of the bitumen from the surrounding sand, there could be little forward movement on the development of the tar sands without settling the land issues. What transpired was a series of land grabs wherein control of the Athabasca territories moved from the hands of the First Nations who had occupied these lands for countless generations, to the federal government, then later to the provincial government, and finally into the hands of the big and increasingly foreign-owned oil companies.

The traditional claim of indigenous peoples to the Athabasca lands is rooted in Aboriginal rights that were enshrined in British and, eventually, Canadian law. Essentially, indigenous peoples who have occupied their traditional lands "since time immemorial" have property rights, which they retain unless they are extinguished through a negotiated settlement.[14] These provisions were outlined in the Royal Proclamation of 1763, which, in turn, was incorporated in the British North America Act (BNA). The original intent of the Proclamation was to protect tracts of traditional lands for indigenous peoples so that they could exercise their hunting, trapping and fishing rights without interference from the early white settlers. When Ottawa repatriated the constitution in 1982, the core elements of the BNA Act, including these rights, were incorporated into the new Canadian Constitution, and they have been upheld by the courts in the face of numerous challenges since.

Shortly after Confederation, Canada made moves to secure control over these vast tracts of traditional "Indian Lands," as they were called, which had been considered worthless until they were found to contain gold and oil. Between 1871 and 1923, the federal government engaged in a treaty-making process with First Nations in various regions of the country. In exchange for giving up their traditional rights to these lands, Canada would provide each indigenous person who signed a treaty the paltry sum of five dollars a year plus one square mile of property for each family of five, along with incidentals such as some twine and ammunition. In 1899, the indigenous peoples of Fort Chipewyan and other First Nations communities in the Athabasca region signed one of theses treaties, Treaty 8, which encompassed the greater portion of northern Alberta plus a part of the Northwest Territories. By doing so, Ottawa claimed, the indigenous peoples had effectively "ceded, extinguished and surrendered" their traditional rights to this land — a position held to this day.

Later, however, this particular treaty-making process was shown to be fraudulent. Subsequent research disclosed that in signing Treaty 8, "the Indian people did not understand or agree to the terms appearing in the written version of the treaty."[15] The purpose of the treaty and its provisions were not sufficiently explained to them in their own languages. White missionaries travelled with Canada's treaty party, translating and explaining the deal to First Nations leaders. Elders who recall the oral version of Treaty 8 and the other treaties negotiated during this period remember them as "peace and friendship treaties," not land surrender deals. Also, most of the seventy-five signatures on Treaty 8 (X marks) are forgeries, shown by handwriting analysis to have been made by the same person. Almost a century later, in 1973, recognizing that the rush for gold, diamonds and oil was heating up, the Indian Brotherhood of the Northwest Territories used this and related evidence to initiate court action and file a caveat on their traditional lands, including the Athabasca region. After studying the evidence and hearing from witnesses in fifteen communities, Mr. Justice William Morrow of the Northwest Territories Supreme Court ruled that the Dene people had a right to file claims and place a caveat on their traditional lands.[16] To this day some First Nations are claiming this so-called Crown land as their own, thereby challenging Canada's right to open it for massive resource projects without their consent.

Although the Athabasca tar sands were not the focal point of this court ruling, they were part of the original Treaty 8 land grab by Ottawa. At that time, Alberta did not even exist as a province and the tar sands were under federal jurisdiction. The early settlers in what is now Alberta, anxious to get their hands on the rich resources of the region, lobbied hard for provincial status; they finally got their wish in 1905. However, Ottawa refused to transfer control of natural resources to the new province. At the time, the rationale was that the people of Ontario and Quebec had basically put the new province of Alberta on its feet by purchasing the land from the Hudson's Bay Company, surveying the territory and drawing up boundaries, and they invested in the region and encouraged settlers to go west. Central Canada wanted a payback and so insisted that Alberta's resources remain in Ottawa's hands for the time being.

It was not until 1930 that Ottawa finally agreed to hand over mineral rights to Alberta, giving the province control over the coveted tar sands region covering around 140,000 square kilometres. However, the federal government retained control over approximately 5,100 square kilometres of prime tar sands land.[17] Although this was only a small portion of the tar sands area, it was the most easily exploited because the bitumen lay close to the surface. As it later turned out, Ottawa had its own plans to showcase the development of the tar sands, starting with this choice piece of real estate. Meanwhile, the new province had other challenges to worry about, including the spread of drought and the onslaught of the Great Depression. A relatively poor province such as Alberta, Ottawa officials reasoned, would not be in the position to develop the tar sands itself.

During the 1930s, federal officials set the stage for the initial development of tar sands oil. American oil producer and engineer Max W. Ball, of Denver, was persuaded by Sidney Ells to invest in the tar sands through the creation of Abasand Oils Ltd. By 1940 Abasand Oils was engaged in extracting and processing operations near Fort McMurray. Worried about wartime oil supply shortages, the government of Mackenzie King increased investment in developing the tar sands. In June 1942, C.D. Howe, the powerful minister of Munitions and Supply in the King government, after getting an assessment of the capacity of the Abasands plant from the Consolidated Mining and Smelting Company, announced that Ottawa would be taking it over in order to provide oil needed for the war effort. Although Alberta reluctantly agreed to collaborate with Ottawa on this venture, it was not evident that Abasands had fully mastered the oil sands.[18,19] Shortly thereafter, the discovery of a rich body of bituminous sands was made in the Mildred and Ruth Lakes area, the area that is currently being developed by Suncor, one of the largest and most active corporate players in the tar sands today.

Ottawa's experiment with the Abasand project never really got off the ground. Plagued by two mysterious fires in the 1940s, Abasand's extraction plant near Fort McMurray sputtered along amidst a series of setbacks. Preoccupied with the fallout from the Great Depression plus an increasingly demanding war in Europe, the federal government was unable to maintain the kind of oversight required by its Abasand pilot project. During the war, Abasand was only producing 17,000 barrels a year. By the end of the war, Canada was still importing close to 90 percent of its oil, mainly from the US. The price of oil dropped as a wave of new conventional oil wells came into production around the world. Facing bankruptcy itself, the Alberta government foresaw its own oil future diminishing further as the Turner Valley oil fields declined and one company after another was drilling dry wells in the Alberta oil patch. Then, in 1947, Imperial Oil struck oil at Leduc, just south of Edmonton, doubling Canada's oil reserves and igniting a boom in Alberta. The number of oil-producing wells jumped from 523 to 7,400 over the next ten years or so.[20] Still, Alberta's resentment of Ottawa's seizure of the small patch of prime real estate in the tar sands continued to fester.

Indeed, the Manning government in Alberta had already called for a royal commission inquiry into the federal government's operations in the tar sands and what it called a "wanton plunder of provincial rights."[21] Ottawa was accused of sabotaging and discrediting the prospects of large-scale development of Alberta's tar sands. When the Alberta government's request for a royal commission was refused, Karl Clark and others urged the province to go ahead and try to develop a successful tar sands project where Ottawa had failed. But the Manning administration was more interested in getting the private sector to take the lead. To show private investors the feasibility of commercially developing the tar sands, the Manning government built and operated a pilot extraction plant at Bitumount. Then Clark's long-time assistant, Sidney Blair, was appointed by the Alberta government to develop a comprehensive report on the economics of producing oil from the tar sands. With considerable fanfare, Blair's report, describing how a profitable operation capable of producing 20,000 barrels of oil a day could be developed in the tar sands, was released to the public in 1950.

Bolstered by the Blair report, Premier Manning convened the first Oil Sands Conference in September 1951. The four-day event brought the leading petroleum and mining companies on the continent to the University of Alberta in Edmonton to discuss how to take advantage of one of the world's greatest untapped resources. The conference was organized to showcase the potential of the Alberta tar sands for substantial oil production, the scientific and technological breakthroughs made, and the scheme's commercial viability. At that point, says author Larry Pratt, "after four decades of frustrating, lonely pioneering work in the wilderness of northeast Alberta" by scientists in both the federal and provincial governments, the decision was essentially made "to place the vast bituminous sands under the monopoly control of the globe's biggest resource extraction companies.... [and] ... turn over forty years of publicly financed research and technical know-how" to the private sector.[22]

As an incentive, the Manning government offered the big oil companies and their investors a uniquely generous leasing deal for land in the tar sands. Prospecting permits and leases would be issued for sites of up to 50,000 acres (20,250 hectares) on a first-come, first-serve basis.[23] The permits for prospecting would be granted for one year, but renewable for the next two years, for a fee of 5 cents an acre for the first year, increasing to 10 and then to 25 cents over the following two years. Once permit holders showed evidence of having conducted explorations on their acreages, they were eligible to lease their land tracts for twenty-one years. The lease would give the company the exclusive right to develop the tar sands and produce crude oil on the land for an annual fee of a dollar per acre. This fee was later reduced to 25 cents per acre. The leases were also renewable for additional twenty-one-year terms. By most standards, the royalty payments to the Alberta government would be minimal, no more than 10 percent of the value of the bitumen extracted. Moreover, Manning's minister of Mines and Minerals made it clear that these leases were available to the private sector only, not other governments or their agencies.

A decade after the Alberta government's tar sands leasing program was announced, most of the prime acreage had come under the control of the major oil companies. In 1958, Imperial Oil and smaller companies like Cities Service and Richfield Oil banded together to lease some two million acres (one million hectares) in the tar sands.[24] By the early 1960s, several major companies were making strip-mining plans while Shell Oil had plans for constructing an in-situ operation. But, for the most part, the larger oil corporations were content to buy up the leases and keep the resource out of production until market conditions were more favourable for profitable exploitation of the tar sands. By taking up the exclusive and cheap concessionary rights to the prime tracts of land, afforded by the Alberta twenty-one-year renewable leasing scheme, they could bide their time while keeping out potential competitors. Had either the federal or provincial government (or both) decided to develop the tar sands itself instead, chances are crude oil would have been flowing from the Athabasca region sooner.

Another damper on the development of the tar sands was that the 1960s was a period of worldwide oil over-production. Already, Alberta was encountering major difficulties in finding markets for its surplus production of conventional oil. One option, initially outlined by Ottawa in its original 1961 National Oil Policy, was to open up the Montreal market, which was being supplied by US oil producers, to Alberta production.[25] The plan was to construct a trans-Canada pipeline that would link the Alberta oil patch to markets in Montreal and the eastern provinces. However, Washington and the US oil majors were bitterly opposed, the plan was dropped, and Alberta's oil producers were forced to find markets elsewhere. For the tar sands industry, the door was effectively closed for the time being. Under the Alberta government's new guidelines, developed after the National Oil Policy was revised, tar sands producers were required to secure markets not already served by the conventional oil producers before they could submit applications for development.

Nevertheless, the "world's first oil mine" was launched in September 1967. The dedication ceremonies for what was called the Great Canadian Oil Sands took place on the Athabasca's west bank just north of Fort McMurray with 500 dignitaries present. The Great Canadian Oil Sands was the brainchild of John Howard Pew, chairman and patriarch of Sun Oil of Philadelphia. Speaking to the assembled audience at the launch, Pew said: "It is the considered opinion of our group that if the North American continent is to produce the oil to meet its requirements in the years ahead, oil from the Athabasca area must of necessity play an important role."[26] In other words, Pew saw the tar sands as North America's oil insurance policy. This pioneering plant, however, did not fare well. Cost overruns gave rise to heavy losses: the company did not show a profit until the first quarter of 1974. These problems, in turn, provided other tar sands companies with the ammunition they needed to negotiate better terms with Alberta. As well, the project became a tax burden on the province, as the government paid millions of dollars in royalties to the company.

It was Syncrude that largely set the terms and conditions for the tar sands industry. At the time, Syncrude was a consortium of four US-controlled oil companies, Exxon (Imperial Oil), Gulf, Atlantic Richfield, and Cities Service. In the early 1970s, Syncrude mounted a powerful lobbying machine that influenced decisions at the highest levels of government in Alberta and Canada over vital energy decisions concerning pricing, taxation, environmental regulation, labour legislation and other policies affecting the development of the tar sands industry.[27] Moreover, the Syncrude consortium managed to convince both the federal and provincial governments to provide the burgeoning tar sands industry with generous, publicly funded incentives in the form of subsidies, thereby underwriting much of the risk and costs associated with production, in return for little in the way of actual ownership and control. In 1971, on its land lease No. 17 at Mildred Lake, which was located next door to the Great Canadian Oil Sands, Syncrude began producing 125,000 barrels per day.

By the early 1970s other players in the oil industry were lining up to put their land leases into production. Shell Canada Ltd., which held land lease No. 13, located across the Athabasca River from the Syncrude site, planned to put its strip-mining and extraction plant into operation. North of the Shell site, a block of leases on Daphne Island had been designated for the Athabasca Oil Sands project, a consortium composed of Canadian subsidiaries of both European (Petrofina SA of Belgium) and US oil companies (e.g., Phillips Petroleum and Continental Oil). The other major mining and extraction operation, expected to produce 103,000 barrels a day, was to be developed just south of the Shell site by Home Oil, then the largest Canadian oil company. By the mid-1970s, however, it was clear that the tar sands had ended up mainly in the hands of US and multinational oil companies.[28]

There was one attempted land grab in the 1970s, however, that did not succeed. A consortium of oil companies, led by Imperial Oil, planned to construct a pipeline corridor called the Mackenzie

Valley Pipeline to transport natural gas from the Beaufort Sea south to markets in Canada (including the tar sands) and in the US. The proposed pipeline corridor would pass through disputed land claimed by the Dene Nation in the Northwest Territories (NWT). In Ottawa, the Liberal minority government of the day with the New Democrats holding the balance of power, appointed Mr. Justice Tom Berger of the British Columbia Supreme Court to conduct a Royal Commission Inquiry into whether the pipeline should be built, paying particular attention to indigenous land rights issues. Following Judge Morrow's 1973 ruling, the Dene Nation had filed a caveat on the land in question. After holding public hearings in dozens of northern communities, and hearing technical evidence in Yellowknife, Justice Berger called on the government to put a moratorium on the construction of the Mackenzie Valley Pipeline until there had been a just settlement of the land claims of the Dene and related environmental concerns were satisfactorily addressed. Berger's recommendations were generally adopted by the federal government in 1977 and land claims negotiations were conducted over the following decade with several First Nations in the NWT. Yet, as we shall see later, the land struggle was never fully resolved.

Petro-Politics

The 1970s was not just the time when the tar sands were being handed over to large, and largely foreign, corporations, it was also an era of turbulent petro-politics. It began with the oil crisis of 1973 when the Organization of Petroleum Exporting Countries (OPEC) slapped an embargo on western supplies of petroleum because of the US and their allies' support for Israel in the Yom Kippur War. As the embargo tightened, oil and gas was rationed in many parts of North America, igniting a public outcry against the major oil companies who profited as a result. In Canada this period of petro-politics was largely dominated by two political titans: Alberta's Premier Peter Lougheed and Prime Minister Pierre Trudeau. Battle lines were drawn between Alberta and the Government of Canada, not only over control of the oil patch, but also about what constitutes a made-in-Canada energy policy, one that serves both the energy security needs and the environmental priorities of people in this country. The battles fought culminated in the rise and fall of the infamous National Energy Program. All along, the tar sands were the major prize at stake in the battle plans that unfolded.

As it turned out, 1973 was a pivotal year for Canada. Shortly before the OPEC oil shock, the Alberta government reached a landmark agreement with the tar sands industry. On September 18, 1973, Premier Lougheed went on province-wide television to announce that a deal had been reached with the Syncrude consortium.[29] After ten years in the planning, Syncrude's billion-dollar project had been given the green light. In the premier's view, the Syncrude complex of mining, extraction and upgrading would provide a model of how the tar sands could be efficiently and profitably developed. By tapping into one of the world's largest reservoirs of hydrocarbons, the Syncrude project would be supplying export markets mainly in the US while creating thousands of jobs for Albertans. The announcement was designed to send a clear signal to Ottawa that Alberta was in charge of tar sands development, and to the oil industry that the tar sands was open for business. (As we shall see later, Lougheed has recently modified his position and is now more critical of the tar sands regarding both their environmental effects and the royalty regime that governs them.)

The shockwaves from the OPEC oil embargo of 1973 were a mixed blessing for Canada. The cartel's decision to cut oil production by 5 percent every month, for as long as western countries ignored their demands about the Middle East Crisis, sent energy prices soaring. Quebec and the Atlantic provinces were completely dependent on foreign oil. Although Alberta benefited considerably from rising energy prices through their oil exports to the US, the eastern part of the country suffered from the escalating costs of imports from countries such as Saudi Arabia. The spike in oil prices did make crude oil production from the Alberta tar sands more economically viable but the reliance on US companies for oil shipments to eastern Canada also raised questions about energy security, especially in situations where the US was in dire need of that oil. As a result, Ottawa began to put priority on safeguarding against such oil shocks in the future by changing the orientation of Canada's energy policies from continental to national.

In December 1973, Prime Minister Trudeau outlined his government's new energy policy. The stated goal of the policy was to achieve "Canadian self-sufficiency in oil and oil products" by the end of the decade.[30] A national oil company, Petro-Canada, would be created to facilitate this transition, and a pipeline from Sarnia, Ontario, to Montreal would be constructed to move western Canadian oil to the eastern provinces. At the same time, the prime minister announced that crude oil exports would be reduced by at least 10 percent and export taxes applied. The federal government would also attempt to "Canadianize" the oil industry by monitoring US investments more closely through the newly created Foreign Investment Review Agency (FIRA) and the Canadian Development Corporation (CDC). To offset the impact of world oil prices, there would be a temporary freeze on domestic oil prices. Finally, Trudeau signalled Ottawa's renewed interest in the tar sands by pledging 40 million dollars in federal funds to kickstart more research and development and the direct involvement of the new national oil company. "Technologies must be developed which do not exist," he said, "in order to permit the development of 85 percent of the sands which are deeply buried."[31]

Reactions to Ottawa's energy agenda were swift and cold. US

companies such as Sun Oil protested that the proposed screening of future foreign investment by FIRA and CDC was a slap in the face given the time, energy and money they had already invested in the tar sands. Creating and imposing a state-owned company, Petro-Canada, was seen as "an egregious affront" implying that Ottawa had little trust in the oil industry. The decision to freeze domestic oil prices, albeit temporarily, was seen as posing a particular hardship on the operations of Syncrude and the Great Canadian Oil Sands, especially in the hyperinflationary economic environment of the time. Added to this was the injury of Ottawa's announcement that the oil companies could no longer deduct their provincial royalties from their federal taxes. The US State Department reacted negatively to Ottawa's decision to impose export taxes and slash oil exports (which had reached a million barrels a day from Alberta by 1972), declaring that such actions would amount to a breach of the "special relationship" between the two countries. According to the industry, these problems would have to be resolved if the federal government expected the Alberta tar sands to play a major role in Canada's energy future.

Tensions between Alberta and Ottawa heightened. Calling the Trudeau energy policy "the most discriminatory action taken by a federal government against a particular province in the entire history of confederation," Lougheed announced his own package of provincial legislative initiatives designed to counter Ottawa's measures.[32] In addition to asserting Alberta's control over oil production and the pricing of crude, Lougheed's plan focused on incentives to accelerate the pace of development in the tar sands, particularly in-situ production. He announced, for example, the "Energy Breakthrough Project," which would provide a pool of 100 million dollars in capital to stimulate the development of new technologies for mining the tar sands. To guide the research and development needed, the Lougheed government also established the Alberta Oil Sands Technology and Research Authority. Yet, instead of providing an overarching strategy for developing the tar sands, Lougheed tended to introduce temporary, almost piecemeal, measures, not only to stimulate and guide further development but also to counter Ottawa's moves.

Fifteen months later, Atlantic Richfield announced it would be withdrawing from the Syncrude consortium, leaving Imperial Oil, Gulf Canada and Cities Service to find another partner to take over 30 percent of the investment. When no oil company was willing or able to fill the void, the federal and Ontario governments joined Alberta in a rescue mission to save Syncrude. What some observers at the time labelled a "sweetheart deal" was worked out in a Winnipeg hotel room and announced on February 3, 1975. Under the new deal, Syncrude obtained an exemption from paying federal taxes on its provincial royalty payments and was permitted to charge prices for its oil that were closer to world prices. The new deal also allowed Syncrude to sell its full production to the US export market, rather than restricting a portion of it for the domestic market. In short, Syncrude's rescue plan included "huge infusions of public cash" along with significant exemptions from Ottawa's new energy policy.[33]

Less than four years later, another oil shockwave struck, this time as a result of the revolution that was brewing to overthrow the pro-US regime of the Shah of Iran. In December 1978, Iran stopped exporting oil, thereby creating a worldwide shortage and a "crisis of confidence" in the US. By July 1979, then US President Jimmy Carter launched an energy plan that included the creation of the state-sponsored Energy Security Corporation to produce synthetic crude oil from unconventional sources such as coal and shale. The Carter administration was reportedly prepared to commit 88 billion dollars to this "synthetic fuels alternative" plan, which dwarfed anything that the Trudeau government had planned to be doing by this time under its new energy policy.[34]

Although Carter's strategy was largely aimed at breaking the stranglehold that OPEC had on world oil prices, he lost his presidency in 1980, before his plan could be implemented. Back home, the Trudeau minority government also lost power for a brief period in 1979 to Joe Clark's Progressive Conservatives, who were committed to dismantling the new policy, and selling Petro-Canada. But, by the spring of 1980, Trudeau had been swept back into power with a majority government.

Having been elected on a platform of economic nationalism, the Trudeau government brought forward its National Energy Program (NEP) as the centrepiece for the reorganization of the Canadian economy. With Trudeau confidant Marc Lalonde installed as federal energy minister, the NEP provided a detailed blueprint for a made-in-Canada energy policy. Current energy policies, said Lalonde, in his introduction of the NEP, "are no longer compatible with the national interest." Canada's oil industry was dominated by US-based multinational corporations, who owned between 70 and 80 percent of Canadian petroleum firms. In order to change direction, argued Lalonde, the NEP would attempt to achieve the following goals: (1) "establish the basis for Canadians to seize control of their own energy future through *security* of supply and ultimate independence from the world market"; (2) "offer to Canadians, all Canadians, the real *opportunity* to participate in the energy industry in general and the petroleum industry in particular"; and (3) "share in the benefits of industry expansion [by establishing] a petroleum pricing and revenue-sharing regime that recognizes the requirement of *fairness* to all Canadians no matter where they live."[35]

The NEP contained several components that directly affected the tar sands industry.[36] First, a "made-in-Canada" price for tar sands crude and heavy oil was established that was tied to, but pegged below, world oil prices. Second, the "Canada Lands" composed of petroleum-bearing regions outside of Alberta (i.e., the

Arctic, Atlantic and Pacific Oceans) was established along with incentives to lure investment. Third, the Petroleum Incentive Payments program was put in place to encourage Canadian companies to invest in the Canada Lands. Fourth, Petro-Canada would lead the Canadianization process by purchasing some foreign-owned companies and by including a buy-back clause to acquire up to 25 percent ownership of all new oil development projects in the Canada Lands. Fifth, a variety of new taxes were introduced to provide the federal government and its agencies with the revenue needed to carry out the NEP. These taxes included the Petroleum and Gas Revenue Tax on the resource at the wellhead; a "Canadian Ownership Charge" on oil and gas leaving the refinery (in order to pay for Petro-Canada's buy-outs of foreign-owned projects in this country); and a tax on gasoline at the pump to help finance Ottawa's subsidy on imported oil for the eastern provinces.

The NEP struck like a bomb in Alberta. In his televised response, Lougheed declared that Ottawa had "simply walked into our home and occupied the living room," and that Albertans were "entitled to respond in a measured way."[37] Alberta's official reaction, Lougheed announced, would begin with a 15 percent reduction in sales of oil and other petroleum products to the rest of Canada, amounting to 180,000 barrels a day. Recognizing that Ottawa would need oil from the tar sands to replace declining stocks of conventional oil in order to fulfill its NEP objectives, Alberta would put a hold on the development of two new tar sands projects, namely, the Alsands and Imperial Oil plants in Cold Lake. In addition, Alberta would mount a legal challenge to the Trudeau government's natural gas export tax instituted under Ottawa's previous energy policy. For Lougheed, and for Trudeau as well, these moves were viewed as bargaining chips to be used in subsequent negotiations between Alberta and Ottawa over Canada's energy policies. However, they were seen as a potential threat to national security by Lalonde and others in the federal cabinet.[38]

Nevertheless, after the heat simmered down, Ottawa and Alberta began to negotiate agreements in regards to particular components of the NEP. Although settlements were reached on key issues, the tar sands industry was not happy with the deals negotiated nor with the fact that the basic NEP structure remained in place. The higher price agreed on for crude oil produced from the tar sands, for example, was pegged at 85 percent of the world price, which, the industry argued, would still hamper production.[39] The settlement on the NEP's tax regime, complained the tar sands industry, continued to penalize the conventional oil production side of their operations, which provides a large proportion of the start-up capital needed for tar sands projects.[40] What finally shook the tar sands industry was the collapse of the Alsands consortium. According to the companies in Alsands, mostly foreign-owned, the new pricing and tax regimes of the NEP, coupled with a doubling of their start-up costs (due to double-digit inflation and interest rates, and the consequent rising labour and material costs), had made the project too risky. Despite the rise and fall of world oil prices during the first half of the 1980s, the oil industry blamed the NEP for depriving them of the revenues required to develop the tar sands.[41]

Meanwhile, the US petroleum giants, aided and abetted by the newly elected US president Ronald Reagan, mounted a campaign of economic retaliation against the Trudeau government and its NEP.[42] In no uncertain terms, the Reaganites in Washington made it clear that they wanted the NEP "buried" and FIRA "declawed."[43] The "Canadianization" provisions in the NEP, they argued, were anti-American and would have the effect of driving US companies, such as J. Howard Pew's Sun Oil, and their investment capital back across the border. By December 1980, many US companies operating in the Alberta oil patch had packed up and hauled their equipment and their investment capital down south.

As the December 15, 1980, edition of *Oilweek* described the scene, companies had already "stacked" 200 oil rigs and were prepared to leave Canada altogether. At the time, the deregulation measures in the US energy industry being instituted by the Reagan administration made the US a more lucrative place to invest. In effect, both the oil industry and the Reagan administration had pulled off what amounted to a capital strike against Canada.

Despite the National Energy Board's 1980 prediction that two to five new tar sands plants would be in production by 1995, the industry doubted that a single project would get off the ground unless the tar sands were made more economically viable. In a 1982 report, the Canadian Petroleum Association (CPA) outlined several measures that the federal and Alberta governments would need to undertake in order to revitalize the tar sands industry.[44] Emphasizing that each new tar sands plant requires a large initial outlay of capital, the CPA urged that the tar sands industry be allowed to charge world prices for the crude they produced. To avoid conflicts with conventional oil producers, the CPA argued that the companies engaged in the production of crude oil from the tar sands should be able to develop their own markets. To do so, the CPA called on the federal government to revise its export quotas to allow producers to sell tar sands crude to the US and other markets.

By August 1981, however, the Trudeau government signalled that it was bowing to the massive US pressure that had been mobilized against the NEP. At that time, finance minister Allan MacEachen announced in his budget speech that the NEP would not be used as a model for the Canadianization of other industries, as originally planned. Nor would legislative action be taken to expand the powers of FIRA to monitor and control foreign investment. Tensions subsided further when Trudeau's 1982 cabinet shuffle sent the embattled Lalonde from energy to replace MacEachen as finance minister.

Shortly after taking over the finance portfolio, Lalonde met with big-business leaders and allegedly made a peace pact with them, promising the government's cooperation and support in the future. Recognizing this shift in the Trudeau government's strategy, the Business Council on National Issues (BCNI), which represents the chief executive officers of the 150 largest corporations in Canada, started to weigh in.

The BCNI set up its own task force on energy policy, composed of industry leaders and chaired by the CEO of Imperial Oil. One-on-one meetings were held with Lougheed, then Ontario premier Bill Davis, and the new federal energy minister, Jean Chrétien. In November 1983, the BCNI organized a two-day summit meeting between key leaders from big business and government behind closed doors at the Niagara Institute. Eight months later, in June 1984, there was a second summit held, except this time government officials participated, for the most part as observers. In other words, the final details of the new energy policy for Canada were hammered out mainly by industry representatives. Unsurprisingly, the energy deal engineered by the BCNI called for the scrapping of the NEP, the adoption of world prices, and an overhaul of corporate taxation.[45] With the oil industry and their big-business allies back in the driver's seat, the days of a made-in-Canada energy policy came to an end. And the black-gold rush saga entered a new phase.

Energy Surrender
The landslide victory of Brian Mulroney's Progressive Conservatives in the 1984 general election marked a turning point in Canada's energy policies. The Trudeau experiment in economic nationalism had come to an abrupt end. So, too, had the attempt to develop a made-in-Canada energy strategy. The Mulroney government's free-trade agenda would soon serve to transfer control over energy policy-making in this country from the state to the

market, where the US petroleum giants reigned supreme. With this dramatic shift in the paradigm for energy policy-making, the tar sands industry finally got the boost it needed to make the eventual breakthrough into boom times. In due course, it would also spark diverse crude awakenings for the country as a whole.

Immediately after assuming power, Mulroney began to remove the chill from Canada–US relations that had come about during the Trudeau era through a rapprochement with the Reagan administration in Washington. Speaking to a gathering of US investment leaders in New York in December 1984, the new prime minister declared that "Canada is open for business." Within the first few months of being in office, the Mulroney government made moves to effectively "declaw" FIRA by removing those provisions that had proven to be most offensive to US investors.[46] Moreover, FIRA itself was dismantled by the Mulroney government in 1985 and was later replaced by Investment Canada, which had the radically different mandate of promoting rather than regulating new foreign investment in the country. Washington was pleased that the new Mulroney government not only fulfilled its campaign promise to scrap the NEP, but also that it made the official announcement to an American business audience.

However, the most significant change in the direction of Canada's energy strategy would come in the negotiation of a free trade deal with the US. Initially, during his campaign for the leadership of the PC party, Mulroney dismissed the idea of a free trade agreement with the US, saying: "Don't talk to me about free trade … Free trade is a threat to Canadian sovereignty." But soon after being sworn in as prime minister, Mulroney took his cabinet to a retreat in the Gatineau Hills with the BCNI.[47] Although the BCNI's policy task forces had several proposals to impress upon the new government, their number one priority was the negotiation of a free trade deal with the US. A few months later, at the "Shamrock Summit" in Quebec City in March 1985, Brian Mulroney and Ronald Reagan announced their commitment to launch free trade talks between the two countries, while singing "When Irish Eyes Are Smiling." The Mulroney government's resolve was further reinforced in September 1985 when the Royal Commission on Economic Union and Development Prospects, which had been appointed by the Trudeau government, recommended free trade with the US. Calling it "a leap of faith," Commission chairman and former Trudeau cabinet minister Donald MacDonald presented his report, which amounted to a expanded version of the blueprint submitted by the BCNI to the Commission.[48]

The Canada–US free trade negotiations were formally launched in May 1986 and concluded in October 1987. Based on his previous experience as a labour negotiator, Mulroney calculated that Canada's oil, natural gas and hydroelectricity could be used as bargaining chips to lure the Americans to the table and seal the deal. The strategy worked. After all, what Washington wanted most from these negotiations was secure access to Canada's storehouse of energy resources, especially oil and natural gas. And Ottawa was only too willing to oblige. For Mulroney, the practice of preserving portions of Canada's oil reserves for Canadians was "odious." Instead, he viewed it as a continental resource to be shared with the US. As the negotiations entered their final week, Mulroney sent finance minister Michael Wilson and trade minister Pat Carney to Washington to close the deal. When the negotiations were concluded, there was no doubt that a continental energy deal was the centrepiece of the new free trade regime.

The core of the so-called energy chapter of the Canada–US free trade agreement (FTA) is the proportional sharing clause.[49] It was designed to ensure that the US would have secure, unfettered access to Canada's energy resources. Known also as Article 904, the proportional sharing rule obligated Canada to provide continuous exports of energy resources to the US. Under this rule, Canada was prohibited from placing a ban or even a quota on its exports of oil and natural gas to the US, even in times of critical domestic shortages, unless Canadian consumption were cut back by a proportional amount. Nor would Canada be allowed to impose taxes on the export or import of oil, natural gas or other energy resources. The rule also ensured that export prices would not be set above the level of domestic prices. In other words, the proportional sharing rule guaranteed that Canada's oil and other energy resources would be treated as continental commodities, destined for US markets and controlled by the US petroleum industry.

At the same time, the National Energy Board (NEB) was stripped of its main regulatory powers over energy exports. Under the FTA, the NEB no longer had the right to "refuse to issue a licence … or revoke or change a licence for the exportation to the United States of energy goods."[50] Nor would the NEB have the regulatory power it once had to approve short-term energy export sales and the NEB's requirement that companies exporting oil and gas file an export impact assessment was dropped as well. The NEB's mandate to ensure energy security, by requiring that there be a twenty-five-year surplus of oil and natural gas before it granted approvals to export applications, was first amended to require only a fifteen-year surplus, and then abandoned altogether as a rule. In effect, the NEB was relegated to the role of monitoring proposed energy projects for their "net public benefit," with little or nothing left in the way of enforcement and regulatory powers.[51]

The proportional sharing rule for energy resources in the FTA would soon provide a major boost to the tar sands industry. The prohibition against the state imposing a ban or quota on the export of tar sands crude, for example, combined with the growing demand for oil in the US, virtually provided the tar sands industry with a guaranteed market for the sale of its products. At the same time, the adoption of the Western Energy Accord, in which Alberta and its neighbours (British Columbia, Saskatchewan and Manitoba) agreed to further eliminate regula-. tory restrictions within their jurisdictions on the export of oil and gas, furthered these free-trade trends. Moreover, under the "national treatment" rules in the FTA, foreign-owned companies, such as the US oil corporations operating in the tar sands, were to be treated the same as (if not better than) Canadian compa- nies, including Petro-Canada. In the mid-1980s, however, world oil prices were relatively low, thereby dampening the impact of the FTA on tar sands production. Nevertheless, the FTA meant that Canada's energy future, and the tar sands in particular, would be driven by free market energy policies. As then former Alberta premier Peter Lougheed put it: "The biggest plus of this [free trade] agreement is that it could preclude a federal govern- ment from bringing in a National Energy Program ever again."[52]

CHAPTER 2

A Triple Crisis

In many ways, Canadians' perceptions of the world we inhabit have changed radically over the last quarter century, a period that witnessed the rise of the tar sands industry, the demise of the National Energy Program and the surrender of Canada's energy sovereignty to the US. For one thing, the world price of oil has skyrocketed, making the tar sands much more profitable, but also making energy extremely costly for people in this country. At the same time, the views of most people in this country about the future of the planet have also shifted significantly. Our level of environmental awareness has changed dramatically, as more and more people have come to the realization that there are imminent threats to the planet such as climate change and the depletion of natural resources. As a result, what is happening in the tar sands today needs to be reviewed through a new set of lenses.

Jerry Mander of the International Forum on Globalization has coined the term "the triple crisis" to describe the dominant ecological and social forces that are now converging to pose a direct threat to the future of the planet and our industrial society as we know it.[53] This triple threat consists of: (1) the peaking of oil along with natural gas, coal and other fossil fuels, bringing an end to cheap energy; (2) the exponential increase of climate chaos due to the buildup of greenhouse gases in the atmosphere; and (3) the loss of other natural resources such as fresh water, forests, fish, wildlife, biodiversity, ocean habitats and fertile soil around the world. The convergence of these threats, says Mander, is "deadly." And, as we shall see, all three are directly related to the operations of our industrial model of society. In this context, the triple crisis concept provides a new set of lenses through which we need to look in order to understand the consequences of the tar sands mega-machine.

Peak Oil

We are running out of fossil fuels, especially oil. What has been the lifeblood of our industrial system for the best part of the last two centuries is a finite, non-renewable resource found within the earth's crust. The hydrocarbon age, during which society has become increasingly dependent on fossil fuels (coal, oil, natural gas) as the prime source of energy, is coming to an end. The key to understanding what is happening to our industrial lifeblood lies in the complex and controversial notion of *peak oil*, the point at which half of all recoverable oil has been extracted, the point at which production levels start to decline.

The concept of peak oil was largely pioneered by American geologist Marion King Hubbert.[54] Born and raised in Texas, Hubbert knew the oil patch. Though trained as a geologist, he was also a physicist and mathematician. Throughout his career, he held a variety of positions including chief geophysicist for Shell Oil, senior geologist with the United States Geological Survey, and professor at Columbia, Stanford and UC Berkeley. As a geologist, Hubbert knew that it took 500 million years for the accumulated reserves of fossil fuels such as the oil and natural gas we use today to form in the earth. As a physicist, he studied how liquids meander through porous rocks in the earth's crust, so as to better understand the recoverability of oil and gas reserves around the world. As a mathematician, Hubbert was able to ana-lyze oil consumption and supply trends.

Using these tools in 1956 Hubbert set out to determine how long oil production in the United States would continue to increase. He estimated the amount of recoverable oil in the lower forty-eight states based on the most recent geological surveys. Then he stud-ied production, noting that it starts slowly at any newly tapped reserve, then increases rapidly as the most easily accessed oil is taken, until a point is reached at which the remaining oil is harder to get at and production begins to decline. He found that the pro-duction peak usually occurs when half of the total reservoir is extracted. Beyond that point it is either too expensive or too phys-ically difficult to extract more of the oil. On average, between 30 and 50 percent of the oil in a reservoir is recoverable. Hubbert also examined the history of discovery in the United States and found that more new oil fields were discovered in the 1930s than in any other decade before or since (at least up until the 1950s, when he was compiling this research) — even though investment in explo-ration increased dramatically in the decades following the 1930s. Then, after estimating the amount of recoverable oil the US had in 1859 when it was first produced commercially, and tracking the consumption rates since then, which had risen dramatically over the previous fifteen years, Hubbert calculated that the US's oil reserves would peak in 1970. And once the peak was reached, pro-duction would decline. At the time he made his prediction, the US was awash in oil and was the world's largest exporter, so Hubbert and his theories were roundly ridiculed. Yet, as it turned out, US oil reserves did peak at the end of 1970.

Hubbert also used his formula to predict when global oil reserves would peak, but his prediction — that world oil reserves would peak around the year 2000 — was just a little off the mark. Worldwide oil production continued to rise after 2000, in part because of expanding tar sands production, which Hubbert had not factored into his calculations because at the time production from unconventional sources such as the tar sands was not economically feasible. Nevertheless, his predictions may not have been that far off the mark. Since Hubbert's death in 1989, several prominent petroleum geologists have applied their own versions of his method in their predictions of peak oil.[55] For example, Colin J. Campbell, who worked as an exploration geologist for Texaco and Amoco, and subsequently founded the Association for the Study of Peak Oil (ASPO), wrote an influential article, "The End of Cheap Oil," published in *Scientific American* in March 1998, in which he concluded that worldwide oil reserves would begin to decline before 2010, possibly as early as 2008.

In 2005, a team of oil geologists led by Robert L. Hirsch prepared a report on "Peaking of World Oil Production" for the US Department of Energy.[56] According to a survey carried out by Hirsch, predictions about when peak oil is likely to occur vary widely, from 2006 to 2025 and even beyond.[57] However, poor quality and possible political bias in the reported data on oil reserves around the world makes it uncertain as to when the peak will actually occur. As other observers have pointed out, that OPEC sets quotas for oil production as a proportion of "proven" reserves provides an incentive to member nations to falsify their official reserve levels.[58] And if one does, this can have a triggering effect. In 1988, for example, when Venezuela suddenly doubled its official reserve levels, Iran, Iraq and the United Arab Emirates immediately boosted their reserves by two to three times. And, two years later, Saudi Arabia raised its reserve estimates by 50 percent. These boosted levels are still the basis for the official reserve estimates for these countries. Moreover, internal records from Kuwait show that its oil reserves are only half what has been officially reported. And in 2004 Shell Oil was found to be overestimating its "proven" reserves by four billion barrels.[59] Uncertainties about worldwide peak oil conditions are further magnified by the changing prospects of developing hard-to-get-at oil deposits in remote regions of Africa, the Arctic, the Antarctic and offshore at huge environmental costs, given new technologies and soaring oil prices.

Meanwhile, demand for oil is rapidly rising while supplies are becoming increasingly tight. Today, global consumption rates are dramatically affected by the growing industrialization taking place in China and India, the two most populous countries on the planet. According to the International Energy Agency's (IEA) 2007 World Energy Outlook, the demand for oil will increase 37 percent over the next two decades. Worldwide demand will rise from current levels of 84 million to 116 million barrels a day by 2030. Half of this demand will come from the industrializing powers of China and India. If, however, the predictions about being on the verge of a peak in global oil production are accurate, the chance of ever reaching production targets of 116 million barrels a day is simply a pipedream. As well, when demand for oil rapidly rises while supplies are tight, then the oil shocks are likely to be more frequent, longer lasting and more devastating. As former US Assistant Energy Secretary Joe Romm stated: "It is widely agreed that we have entered a new era where the oil market has lost most of its flexibility and capacity to handle disruptions to world oil supplies."[60] In the past, increased oil demands and corresponding supply shocks have been offset and world price swings stabilized by the "spare capacity of OPEC," primarily Saudi Arabia. As recently as 2002, spare capacity was larger than worldwide oil consumption by 10 percent. By 2006, spare capacity had dropped to less than 2 percent.[61]

In addition, new discoveries of conventional oil reserves in the world peaked back in the 1960s and have declined substantially since that time. Up until that time, the two single largest oil fields ever discovered were the Burgan field in Kuwait in 1938 and the Ghawar field in Saudi Arabia in 1948. Appropriately labelled as "elephant" or "super giant" oil fields in the petroleum industry, the Hirsch report described them as the "easiest to find, the most economic to develop, and the longest lived." But, the last "super giants" or "elephants" to be found were in 1967 and 1968. One of these "super giants," the North Sea fields, peaked in 2000 at 6 million barrels a day and is now in sharp decline. By 2010, the North Sea production is expected to have dropped off to 3.4 million barrels a day.[62] Indeed, of the 65 oil-producing countries in the world today, 54 have already passed their peak levels of production and are in a state of "continuous decline."[63] Moreover, given the intense level of oil exploration activity that has gone on since the 1960s, it is highly unlikely another "super giant" conventional oil field will be found.

This brings us to the question of sources for unconventional oil production such as the tar sands, which are the hard-to-get-at and more expensive to extract petroleum reserves. Besides the tar sands, unconventional oil reserves include the extra-heavy oil deposits in Venezuela, the shale oil resources in the United States, the recently estimated oil deposits in deep waters beneath the permanent ice of the the Arctic Circle,[64] and deep sea oil pockets off the coasts of countries such as Mexico and Brazil. Developing these hydrocarbon deposits will be technologically challenging as well as costly. Even if these unconventional sources are fully exploited, there is no evidence that they will be sufficient to reverse the worldwide decline in production following peak oil. According to the IEA scenario, as we have seen, global oil production will have to reach 116 million barrels a day in order to meet world demand by 2030. To reach these targets, says the IEA, unconventional oil production will have to account for 37 million barrels a day. Observers point out that even under the most optimistic scenarios, the Alberta tar sands may be capable of producing 5 to 6 million barrels a day by 2030, and Venezuela's heavy oil production may contribute another 6 million barrels daily.[65] But that leaves a shortfall of 25 million barrels a day, which cannot possibly be met by production from shale oil and deep sea deposits, including the latest oil reserve estimates for the Arctic Circle.

Figure 2.1, prepared by the Association for the Study of Peak Oil (ASPO), portrays historical world trends in oil discoveries and consumption, and projections for the future.

Matthew Simmons, founder of Simmons & Company International, an independent banking firm that specializes in the energy

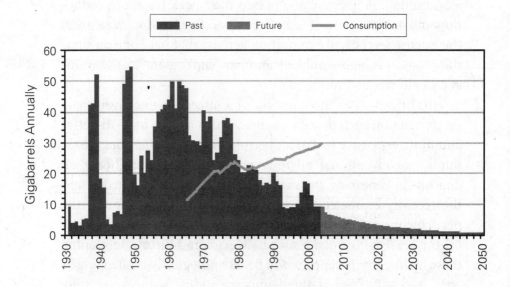

Figure 2.1. World Trends in Oil Discovery and Consumption, 1930–2030.
Source: Association for the Study of Peak Oil.

industry, insists it is time to come to grips with the realities of peak oil. As a lifelong Republican with thirty years of experience in investment banking, Simmons has also been an advisor on energy policy to US president George W. Bush and his administration. In a lecture delivered to the American Association of Petroleum Geologists in June 2001, Simmons declared: "A simple check of the facts reveals that almost every scrap of spare energy [production] capacity around the globe is now either gone or just about to disappear … Even the Middle East is now beginning to experience, for the first time ever, how hard it is to grow production once giant fields roll over and begin to decline." He went on to say there is increasing evidence "that almost every giant field in the Middle East has already passed its peak production."[66] Nor is he counting on tar sands, shale oil and agro-fuel production to make up the shortfall. Current and near-term production from these sources, he said, in October 2007, would be "tiny amounts" and "too inconsequential" to make a major difference in the outcome.[67]

Perhaps it is not surprising that there is little in the way of public discussion about the prospects of a worldwide peak in oil production. After all, stirring up public debate on this issues would likely unleash a wave of panic not only about our energy but also about our economic and social future, given our dependence on a secure, steady supply of oil. The petroleum industry's own spin doctors work overtime to quell public discussion about oil depletion. In fact, as former oil company geologist Colin J. Campbell explains, the very notion of oil depletion is considered taboo in the industry:

> The one they don't like to talk about is depletion.
> That smells in the investment community, who
> are always looking for good news and the image,
> and it's not very easy for them to explain all these

rather complicated things, nor indeed do they have any motive or responsibility to do so. It's not their job to look after the future of the world.... Their directors are in the business to make money for them, so they shy away from the subject, they don't like to talk about it. But they themselves understand the situation as clearly as I do. If they had such great faith in growing production for years to come, why did they not invest in new refineries? There have been no new refineries in the United States for more than 10 years. Why are they always merging? They merge because there is not room for them all. It's a contracting business.[68]

When global oil production does finally "peak," say some observers, it is likely to manifest itself during a period of market instability: a volatile period in which escalating prices result in a recession that dampens demand, and then prices. To some degree, these conditions of market instability obtained during the first half of 2008. In January 2008, the Canadian Imperial Bank of Commerce's (CIBC's) chief world analyst, Jeff Rubin, declared that peak oil had arrived and that oil prices would continue to soar because existing production levels could not meet the rising demand and there are bound to be delays in getting new megaprojects for production from unconventional sources such as the tar sands on stream. As Rubin put it: "Don't think of today's prices as a spike. Don't think of them as a temporary aberration. Think of them as the beginning of a new era."[69]

In any case, when the peak is reached, the Hirsch report warns, its effects will *not* be temporary.[70] There will be dramatic spikes in oil prices, which will hit the transportation sector hardest. Globally, developing countries will be the hardest hit. Greater energy efficiency alone will not be sufficient, says the Hirsch report, to overcome these problems. Furthermore, the economic and social consequences of peak oil and its aftermath will be chaotic. The Hirsch report concludes that nothing short of government intervention will be required. In any case, what is clear is that the peak will only become obvious to most people when the terminal decline begins.

In 2004, when world oil prices began their upward climb, *National Geographic* proclaimed the moment as "The End of Cheap Oil" in one of its cover stories. While analysts in the business press cited events such as strikes in Norway, political unrest in Nigeria and pipeline sabotage in Iraq as causes of the price spike, other energy specialists, such as Richard Heinberg, pointed out that these oil shocks, even when added together, are not sufficient to explain the dramatic and ongoing price surge.[71] Clearly, as noted above, the limited "spare capacity of OPEC" is having an impact. Without that spare capacity, even a slight dip in supply can send prices soaring, especially in the transportation sector where demand is so inelastic. In 2005, a simulation exercise called "Oil Shockwave" was conducted by a blue ribbon panel of White House and national security officials for the US National Commission on Energy Policy, which demonstrated that a relatively minor disruption in global oil supplies would result in a whopping 177 percent increase in the price of oil.[72] In their hypothetical scenario, they calculated that if three million barrels a day (or less than 4 percent of global supply) were removed from the world supply, the price of oil would leap from the then price of 58 to 161 dollars a barrel.[73]

The ongoing price surge has provided a major boost for the tar sands industry. In 2005, industry officials were saying that they needed to get a return of 23 to 28 dollars a barrel for their tar sands crude in order to make production profitable. At that point, the world price was twice that amount. And global prices have continued to soar ever since, with only the occasional dip. By mid-2008, the world price was hovering around 140 dollars a barrel, well over twice what it was in 2005 and five times what had been projected to be necessary for a profitable return on the production of oil from the tar sands. The steady surge in global oil prices since 2004, coupled with growing evidence that oil production was nearing its peak elsewhere in the world, has transformed the tar sands from a sideline into a highly profitable industry and attractive investment opportunity. Because of the price surge, the tar sands companies have been making a steady stream of announcements of revised production targets, expansion plans for their open-pit mining and in-situ plants and new research and development programs.

But peak oil conditions also pose a special challenge for Canada's energy strategy. Regardless of the debate over when peak oil will occur (if it has not already), the fact remains that oil is a finite resource and the world is rapidly running out of it. In the medium to long term, there is no choice but to make a transition from fossil fuels to an alternative energy source. The tar sands, as we have seen, are one of the few remaining large hydrocarbon deposits containing recoverable oil, but increasing dependence on tar sands crude does not solve the peak oil crisis. The peak oil challenge facing Canada is how to manage its development so as to best facilitate the crucial transition from fossil fuel dependency to sustainable energy alternatives. For Canada, the problem of taking up this kind of challenge is further hampered by the surrender of sovereignty over our energy resources. After all, the tar sands mega-enterprise, currently the engine of not only energy production, but Canada's economy as a whole, is primarily controlled by US markets and industry.

Now, and for decades into the foreseeable future, the peak oil challenge will haunt Canada and its political leadership. How the tar sands are developed and managed in response to the peak oil challenge will, in large measure, determine Canada's destiny in the twenty-first century. Will our remaining oil riches be developed in a way that perpetuates dependency on fossil fuels or will they be used to help make an orderly transition to a future powered by alternative energy? Without regaining control over our energy resources, and the development of the tar sands, we cannot even begin to confront the peak oil challenge.

Climate Change

Equally momentous and equally pressing is the challenge of climate change. Not unlike peak oil, climate change brings a whole new set of conditions in the context of which energy policy-making in general, and tar sands development in particular, must be viewed. Climate change, of course, is a result of the warming of the planet. As volumes of carbon emissions are released into the atmosphere every day, they trap heat in the earth's atmosphere, thereby causing increased global warming. With the heating of the planet comes a dramatic array of changes such as melting glaciers, rising sea levels, widespread and prolonged droughts, the loss of habitat and extreme weather events — all of which have serious impacts both environmentally and socially. Although scientists have been warning us about the dangers of global warming for more than two decades, it has taken Al Gore's documentary film, *An Inconvenient Truth*, to bring climate change into popular culture.[74]

In brief, the story of the climate change challenge dates back to 1988. In June of that year, James Hansen, Director of the NASA Goddard Institute for Space Studies, testified before the US Senate Energy Committee that emissions of carbon dioxide and other greenhouse gases were heating up the planet.[75] Shortly thereafter, the United Nations set up the Intergovernmental Panel on Climate Change (IPCC), eventually composed of more than two thousand scientists from around the world, to examine climate issues and the causes of global warming. Four years later, in 1992, the Earth Summit in Rio de Janeiro cast a global spotlight on the world's mounting ecological crises, including climate change. Five years after that, in December 1997, the Kyoto Protocol was negotiated. Ratified by 169 countries before February 2005, the Kyoto Protocol committed its signatory nations to reducing their greenhouse gas emissions to below 1990 levels by 2012. While many countries pledged to cut their emissions by 5 percent below 1990 levels, the Chrétien government committed Canada to a 6 percent reduction. In 2001, the Bush administration pulled the US out of the Kyoto Protocol, thereby delivering a serious blow to global action on global warming.

As George Monbiot explains in his path-breaking book *Heat*, the two principal greenhouse gases are carbon dioxide and methane. Because the greenhouse gases have an insulating effect on the atmosphere, as they build up, more heat is trapped and the planet's temperature tends to rise. Scientists studying ice cores from the Antarctic have concluded that the levels of carbon dioxide and methane in the atmosphere today are higher than at any other time in the past 650,000 years. The release of carbon dioxide is largely a result of the burning of coal, oil and gas, while the release of methane comes from farms, coal mines and landfill sites. Of these two main greenhouse gases, carbon dioxide has the most impact. Its concentration in the atmosphere is currently 380 parts per million as compared with 280 parts per million some four hundred years ago. Most of this increase has taken place over the last fifty years.[76]

In 2007, the IPCC issued a series of reports containing conclusive evidence that the climate change we are witnessing today is a result of human activity. There is nothing especially new in the science behind the IPCC report. Increasingly over the past two decades, scientists from around the world have been sounding the alarm bells about carbon emissions. What is new is the widespread media attention that was given to the IPCC's reports, from front page coverage in newspapers and lead items on the evening news, to topic-of-the-day on talk radio and cover stories in magazines. Unlike peak oil, which has received very little media attention, new levels of public awareness about global warming have evolved and climate change themes are now being woven into popular culture.

Today most climate scientists agree that the critical threshold for temperature rise on the planet is two degrees Celsius above pre-industrial levels. The Potsdam Institute for Climate Impact calculates that holding global temperatures at two degrees Celsius above pre-industrial levels will "stabilize concentrations of greenhouse gases in the atmosphere at or below the *equivalent* of 440 parts of carbon dioxide per million," which would be sufficient to bring global warming under control.[77] During the twentieth century, the average global temperature increased by 0.6 degrees Celsius, above pre-industrial levels, which, according to the World Meteorological Organization, is "the largest [increase] in any century during the past 1,000 years."[78] However, if temperatures rise beyond the two-degree threshold during this century, say climate scientists, then major ecosystems on the planet will begin to collapse. Up to this point, the planet's ecosystems are able to absorb much of the carbon dioxide released into the atmosphere. Beyond this point, the planet's ecosystems start releasing greenhouse gases instead of absorbing them, such as in the case of the melting of the permafrost in the Arctic, which will result in the release of methane gas. At this moment and beyond, says Monbiot, "climate change is out of our hands: it will accelerate without our help."[79]

A 1.4 degree increase over current temperatures is not a "safe" level of warming, says Monbiot. "It is merely less dangerous than what lies beyond." Studies conducted at the UK Meteorological Office estimate that a rise of 2.1 degrees could mean up to 3 billion more people on the planet will face water shortages.[80] The melting of the glaciers in the Himalayas and the Andes pose dangers for people who depend on them for their water supply, especially in Pakistan, parts of China and elsewhere in Central Asia in the case of the Himalayas, and in Peru, Ecuador and Bolivia in the case of the Andes. As ice sheets melt and sea water rises, many coastal cities will be endangered by salt water floods that will not only destroy infrastructure but contaminate freshwater sources. A study published by the Royal Society in 2005 shows that when levels of CO_2 increase, plants give off less water from their leaves, thereby reducing local rainfall and, by extension, lowering crop production in many parts of the world. As well, rising temperatures are bound to contribute to an increase in cholera, diarrhea and malaria. According to one study, a rise of 2.3 degrees Celsius above current levels would expose between 180 and 230 million people to the risk of malaria.[81]

Even smaller increases will have major environmental and human impacts. At a 2005 conference of scientists hosted by the World Meteorological Organization in the UK, an examination of the effect on human beings of rising temperatures produced some startling results.[82] Temperature increases of less than one degree above pre-industrial levels, for example, would trigger a decline in crop yields, the spread of drought within the Saharan region of Africa, a further deterioration of water quality and a dying off of coral reefs. If temperatures rise by less than 1.5 degrees above pre-industrial levels, we can expect that an additional 400 million people will face water shortages, an additional 5 million people will go hungry, and 18 percent of the world's species will become extinct. As well, the complete melting of the Greenland ice sheets will have been triggered. What's more, scientists predict that if substantial reductions in carbon emissions are not made soon, rising temperatures will reach the two-degree threshold by as early as 2030.

In July 2006, James Hansen warned that the melting of the ice sheets of Greenland and the Antarctica would have major destabilizing consequences. A 2.8 degree Celsius increase above current temperatures could well result in a rise of sea levels by up to twenty-four metres, which would put most coastal cities under water. The last time, he said, that the earth's temperature reached this high was three million years ago. Indeed, the warmest interglacial period of the last half million years was when the earth's temperature was just over one degree warmer than it is today. At that time, the sea level rose by almost five metres, which would result in significant damage if it were to occur today. Based on his diagnosis, Hansen declares: "We have at most ten years — not ten years to decide upon action but ten years to alter fundamentally the trajectory of greenhouse gas emissions," in order to prevent these disasters from happening.[83]

The Alberta tar sands are also a key factor in this climate change challenge. As is discussed in more detail in Chapter 5, the tar sands are on the verge of becoming Canada's biggest emitter of greenhouse gases. If current plans proceed for the development of the tar sands between now and 2030, the impacts on the Arctic alone could be devastating. The sea ice in the Arctic has already shrunk to the smallest size ever recorded. A temperature rise of two degrees Celsius would be enough to melt the Arctic sea ice for the summer months. While this would open the Northwest Passage to cargo ships, it would do serious damage to the ecosystems in the region and wildlife populations that depend on them. In the meantime, the melting of the permafrost in the Northwest Territories, which has remained frozen since the last Ice Age, will accelerate. As the permafrost melts, it releases methane into the atmosphere on a massive scale, thereby contributing further to the greenhouse effect and global warming. In short, the production of tar sands crude has already helped trigger, and is bound to accelerate, a chain of meltdowns in the far

north and the Arctic with major ecological and social consequences.

Meanwhile, of course, the world's leading oil companies, many
of whom are involved in the tar sands, have been among the most
prominent naysayers of climate change. Take, for example,
ExxonMobil, the world's most profitable corporation, whose subsidiary Imperial Oil is a major player in the tar sands. According
to an analysis of company documents, ExxonMobil funds a variety of lobby organizations, citizens' groups and academic bodies
in a deliberate effort to undermine the scientific claims about climate change.[84] The message is: the science on climate change is
contradictory and the scientists are split; the environmentalists
are alarmists and liars; and governmental action on global warming would endanger the global economy. Among those known to
be funded by ExxonMobil are the Cato Institute, the Heritage
Foundation, TechCentralStation, the Center for the Study of Carbon Dioxide, the National Wetland Coalition, the National
Environmental Policy Institute, and the American Council on
Science and Health. By funding a wide range of groups to deliver
its message, ExxonMobil was able to sow seeds of doubt about the
climate change challenge. These are the same methods that have
been used by the global tobacco corporations in their campaigns
to stop the adoption of anti-smoking laws.

And ExxonMobil is not alone. A year after the IPCC was
formed, some fifty oil, gas, coal, auto and chemical manufacturing companies came together in the Global Climate Coalition
(GCC).[85] Among the participating oil companies were BP,
Chevron, Exxon, Mobil, Shell and Texaco, most of whom are currently involved in the tar sands industry. Assisted by public
relations giant Burston-Marsteller, the GCC waged an ongoing
campaign warning the public that reductions in greenhouse gas
emissions and restrictions on fossil fuels would essentially ruin
the global economy's promise of prosperity.[86] After the Kyoto

Greenhouse gas emissions billow into the sky above an upgrader and processing facility. Courtesy David Dodge, The Pembina Institute.

Protocol was ratified, a split occurred in the GCC over the issue of climate change. By 2000, several companies had broken away, including some oil companies, namely, BP, Shell and Texaco. With the support of the Pew Center for Global Climate Change, these oil companies joined with Suncor, along with other energy and chemical corporations, and environmental organizations such as Environmental Defense in the US, to form a new coalition called the Partnership for Climate Action. Unlike the GCC, the new coalition accepts the reality of the climate change challenge but advocates market-based solutions.

It was the Stern Report in October 2006 on the economic implications of climate change that reset the parameters of the debate. Sir Nicholas Stern, former chief economist for the World Bank, was commissioned by the government of the UK to examine the

potential costs to the economy of action versus inaction on climate change.[87] If little or nothing is done about global warming,
said Stern, there will be a massive downturn in the annual gross
domestic product of up to 20 percent during the latter decades of
this century. In other words, inaction on global warming would
produce an economic crisis on a scale similar to that of the Great
Depression. (The difference, of course, is that this economic
downturn would be permanent, not cyclical.) Stern went on to
argue that action on climate change now would be much more
economically beneficial than inaction. The costs of inaction, he
concluded, would be five to twenty times greater than the investment needed to take effective action now. It was difficult for most
people in government and industry to dismiss Stern as an
alarmist, given his role with the World Bank and his status
amongst Britain's economic elite.

In Canada the climate change challenge has been largely mis-
handled by successive governments. Although Chrétien's Liberal
government signed the Kyoto Protocol and committed to lowering Canada's greenhouse gas emissions to 6 percent below 1990
levels, this country's greenhouse gas emissions levels rose steadily under Liberal administrations to nearly 30 percent above 1990
levels. The current Liberal leader, Stéphane Dion, acquired a positive international reputation when he displayed ecological
commitment and vision while chairing the UN conference on the
Kyoto Protocol in Montreal in December 2005, but the plan of
action he proposed as then environment minister was considered
by the David Suzuki Foundation and other leading environmental groups to be a "notable setback for climate protection."[88] After
Stephen Harper's Conservatives took over the reins of government in January 2006, Canada's Kyoto targets were dismissed out
of hand. Within a month of assuming office, the Harperites cut
climate change programs, first at Natural Resources Canada and
then at Environment Canada. By the end of 2006, when polls

showed that climate change was the number one concern of Canadians, Harper's approach changed. To create the public impression that his government was prepared to be more active on this file, Harper suddenly switched environment ministers, appointing the combative John Baird, and many of the former Liberal programs on climate change were quickly reinstated. Still, Canada would not be meeting its Kyoto commitment to reduce greenhouse gas emissions.

For Ottawa, the sticking point in dealing with the climate change challenge is the Alberta tar sands. As we shall see in Chapter 5, the burning of natural gas to produce oil from the tar sands produces more than three times the amount of greenhouse gases conventional oil production does. It is quite possible that the Alberta tar sands will soon be responsible for half of all the industrial greenhouse gas emissions in Canada. Yet, as we will discuss further in Chapter 3, the Harper government has close ties with the tar sands industry and has gone to great lengths to protect it from any disruptions that may be caused by the regulation of greenhouse gas emissions.[89] But even if another government were elected, one with a strong commitment to tackling climate change, it would have to face the other challenges. In order to rein in the tar sands industry, Ottawa would have to exercise leadership on matters of energy policy. The national government in this country no longer has full control over energy policy-making, particularly in regards to oil and gas resources. Instead, policy is largely determined by US market forces and petroleum industry interests. Any government committed to reducing the global warming effects of the tar sands industry would have to be prepared to confront this challenge as well.

Nor is it clear that the official opposition is ready and able to give the kind of leadership needed to deal effectively with the climate change challenge. In 2005, when asked what he intended to do as environment minister about the Alberta tar sands,

Stéphane Dion replied: "There is no minister of the environment on earth who can stop this from going forward, because there is too much money in it."[90] Now that he is leader of the Liberal Party, Dion is proposing to tax carbon. His new Green Shift Plan calls for a levy of up to 40 dollars a tonne on fossil fuel emissions, to be offset by personal and corporate tax reductions and monitored by the Auditor General to ensure the tax is revenue neutral. While the Liberals' carbon tax proposal is a bold initiative, there is debate as to whether it is of sufficient magnitude to significantly alter the operations of the tar sands industry and the behaviour of other businesses and of consumers. In Alberta and Saskatchewan, a carbon tax imposed by Ottawa will no doubt be viewed by many as the ghost of the "NEP-brand peril rising again." Meanwhile, British Columbia has introduced its own version of a carbon tax, albeit amidst controversy.

But whether it is a carbon tax or some other measure, bold leadership is urgently needed if Canada is going to measure up to the climate change challenge. Whatever national strategies are undertaken, the tar sands will be of pivotal importance on a global scale. A huge portion of Canada's land mass is in the Arctic and also in close proximity to Greenland. This makes it imperative that Canada develop and implement a visionary and effective plan of action to deal with global warming through mandatory reductions in greenhouse gas emissions. Geologically, we are endowed with perhaps the largest untapped hydrocarbon deposit in the world, which again makes it imperative that creative and decisive action be taken to control and manage the development of this resource for the transition to an alternative, post-carbon energy future. To move in this direction, we will need bold leaders to take back control of Canada's energy policies.

Water Depletion
There is a third crisis threatening the future of industrial society:

the depletion of other natural resources, including fresh water, fertile soils, forests, biodiversity, fish and wildlife, coral reefs, and much more. The abundance of these natural resources has been key to the development of industrial society and to the welfare of humanity in general. Their depletion, at ever increasing rates, poses a growing threat to the survival of people and other species on this planet. Within the next few decades, for example, 50 percent of the world's plant and animal species could be lost.

As we shall see, the development of the tar sands has a serious impact on three of these natural resources, namely, fresh water, the boreal forest ecosystems, and wildlife. However, in order to illustrate this third leg of the triple crisis, our focus will be on the rapid depletion of fresh water on the planet.[91] As we shall see in Chapter 5, water depletion and contamination are also direct consequences of tar sands production.

Of all the natural resources in decline today, fresh water is arguably the most precious. After all, water is the essence of life itself. Nothing exists on this planet — neither humans, nor animals, nor plants — without fresh water. There can be no fertile soils without adequate supplies of water. In a given bioregion, water and forests are inseparably linked, each depending on the other for its existence. Water is the lifeblood of the earth itself. Yet most people living in industrialized societies tend to take water for granted, especially in water-rich countries such as Canada, while freshwater sources are being depleted and contaminated at an alarming rate, not only in water-poor regions of the world but in water-rich regions as well.

As it is with oil, so it is with water. Unbeknownst to most people, industrial society is utterly dependent on large volumes of water for the products we make and consume every day. Throughout the world, agriculture is responsible for 65 percent of human water use, much of this in mass irrigation systems for crop production. Living as we do in a global supermarket, people in Canada and elsewhere depend on food produced all over the world by agribusiness corporations, which threaten the water systems once used by small farmers to produce food for local markets. Many other industries have also become water-intensive. For example, production of the average car in North America (including the steel, rubber, plastic and computer components) uses at least 275,000 litres of water. Computer manufacturing uses massive quantities of de-ionized fresh water in the production of computer chips and related products, often draining aquifers of some of the purest water to be found. Moreover, the mining and petroleum industries use untold quantities of fresh water to extract minerals, oil and natural gas.

As a result, those maps of the world that so vividly portray the disparities between water-poor and water-rich regions, are already undergoing profound changes. According to the United Nations, one-third of the world's population does not have access to supplies of fresh water adequate to meet their daily needs. If current trends continue, by the year 2025 close to two-thirds of people on this planet will be without access to adequate supplies of water. The worldwide demand for water is currently doubling every twenty years, which is twice the rate of global population growth. By the year 2025, say some water scientists, the demand for water will outstrip supply by up to 56 percent. Simply put, this planet is on the verge of a freshwater crisis, which, in turn, runs parallel with the looming crises of peak oil and climate change. As we shall see, these three crises are, to a large extent, interdependent.

Prominent among the multiple causes of the freshwater crisis is the damage that has been done to the planet's hydrological cycle by industrialization. Slovakian hydrological engineer Michal Kravčík and his team of scientists have studied the impacts of urbanization, industrial agriculture and deforestation on the earth's water cycle. Water is a renewable resource because of the hydrological cycle: water falls to the earth in the form of precipitation and then evaporates again into the atmosphere, then falls again. So, if a drop of water, explains Kravčík, lands on a blade of grass, or soil, or the leaf of a tree, or a meadow, or a lake, it continues in the hydrological cycle, evaporating back into the atmosphere. But when that same drop of water falls onto the pavement or buildings of a city, it runs off into the waterways, and eventually ends up in the ocean, and is mixed with salt water.

According to studies conducted by Kravčík and his team, the damage done to the natural flow of the hydrological cycle is depleting the amount of fresh water that is available on the planet. When the earth's surface is paved over as a result of urbanization, says Kravčík, denuding forests and meadows and draining creeks and springs, then natural watersheds are damaged so that rainfall flows out to the oceans faster. Kravčík's team has been able to quantify decreases in renewable water supplies due to increased construction of buildings, paved roads, car parks and additional roofing. Given the current rate of urbanization, the Kravčík team calculates that continental watersheds are losing approximately 1,800 billion cubic metres of fresh water a year. In turn, this causes the oceans to rise five millimetres annually. If this trend continues, by 2025 the world's land mass will have lost 45,000 billion cubic metres of renewable fresh water, which is approximately one-quarter of the volume of water in the entire hydrological cycle.[93]

At the same time, the world's supply of fresh water is also being reduced by the pollution of water systems.[94] The relentless spewing of raw sewage, toxic chemicals and pharmaceuticals into lakes, rivers and groundwater systems has destroyed a substantial portion of the world's remaining freshwater sources. While UN studies show that industrial activity will consume twice as much water in 2025 as it did in 2000, industrial pollution of water systems is expected to have increased four-fold by that year. In the free trade zones or *maquiladoras* of Mexico, for example, manufacturing plants release tens of thousands of tonnes of toxic chemical residue into local waterways every year. Pulp and paper plants dump chemical residue from their chlorine-bleaching process, oxygen-depleting effluent, into waterways, which end up being choked with algae as a result. Factory farms, where hogs, chickens and other animals are crowded into confined spaces for mass production, produce vast volumes of liquid manure laced with antibiotics that are stored in lagoons, until heavy rains, floods or leaks let the material escape to poison groundwater and surface water systems. Meanwhile, the heavy use of nitrogen fertilizers in agriculture, a notorious source of water pollution, runs off into streams that feed river systems. And, increasingly, prescription drugs are finding their way into the waterways, and hormones and other chemicals are turning up in drinking water.

Added to all of this are the impacts of global warming: when the surface temperature of the earth rises, water evaporates from the soil more rapidly with the result that less seeps into the groundwater systems. As well, water in lakes and rivers also evaporates more readily, and, as when the snow packs melt earlier than normal, they tend to evaporate rather than flow into streams and rivers that feed lakes. If, due to warmer temperatures, lakes do not freeze over in the winter, then more of the water is lost to the atmosphere and less is saved in groundwater systems. And once a glacier melts, then the annual spring runoff into rivers and streams is lost, which often results in the drying up of these waterways. Indeed, some water scientists contend that global warming is the main cause of the freshwater shortages looming just over the horizon. Around the world, the water levels in lakes and rivers are expected to drop significantly as a result of climate change. According to some scientists, major portions of the Amazon basin will have become desert by the year 2050.

Here in North America, the looming water crisis is becoming more and more acute. Although both Canada and the United

States are considered to be water-rich countries, endowed with bountiful lakes, rivers and streams, this blue vision of the continent can be misleading. As water scientists point out, a distinction needs to be made between "volume" water and "renewable" water.[95] By "volume" water, they mean the water that is contained in lakes. Lakes themselves have a limited storage capacity and their total volume represents a relatively small percentage of the water required on a year-by-year basis. By "renewable" water supplies, water scientists mean the amount of water flows and groundwater recharge within the borders of a given country. In other words, this is the water that comes from the precipitation (rain, snow) and is collected in rivers and streams or seeps into the ground. Although Canada and the US share joint jurisdiction over the Great Lakes, the largest freshwater body in the world, this is not a renewable water supply, and, besides, the water levels of the Great Lakes have been steadily declining in recent years.

The World Resources Institute (WRI) based in Washington studies the renewable water supplies of countries around the world. To measure renewable water supplies, the Institute calculates the amount of precipitation that falls within the country's borders, which, in effect, recharges the water flows in rivers, streams and groundwater systems of that country. According to the WRI, the two countries with the greatest renewable water supplies in the world are Brazil and Russia. Brazil has 12.4 percent of the world's total renewable water while Russia has 10 percent. Canada comes in third at 6.5 percent, immediately followed by the US, China and Indonesia at 6.4 percent. In effect, the US is virtually in a tie with Canada and three other countries for third place. Yet, for Canada, these numbers are misleading. Approximately 60 percent of this country's rivers flow northward into the northern territories and eventually the Arctic Ocean, away from where the vast majority of Canadians live and work.

As a result, it is estimated that Canada's real portion of the world's freshwater supplies is 2.6 rather than 6.5 percent.[96]

Therefore, the US essentially has two-and-a-half times the amount of renewable water as Canada has, but ten times the population. Not surprisingly, the US today is facing a looming water crisis. To see why, we need to look at how urban and regional demands for water are outstripping local sources and supplies. In the US now, the vast majority of the country's population — almost 80 percent — lives in cities at a time when the watersheds of urban America are being depleted. Surveys currently reveal that in an increasing number of cities, there are signs that traditional water sources are either drying up or becoming so contaminated that new water sources have to be found. According to the Urban Water Council, 24 percent of the US's medium-sized cities and 17.3 percent of its large cities are expected to face serious water shortages by 2015.[97]

The problem of US water shortages becomes even more disturbing when viewed on a region-by-region basis. In particular, there are three major regions of rapidly growing water shortages which, in turn, put additional pressures on Canada to provide freshwater resources to the US.[98]

• *Southwest States:* The fastest growing region in the US is already dry and must pump water in from elsewhere. In Arizona, the city of Tucson has part of its water supply pumped in from the Colorado River, and in the eastern section of Phoenix, which is growing at a rate of almost half a hectare per hour, water tables have reportedly dropped by as much as 120 metres. In California, the water table under the San Joaquin Valley has dropped nearly ten metres in some areas in the past fifty years, while overuse of groundwater in the Central Valley has resulted in a loss of over 40 percent of the water stored in California's reservoirs. Similar trends in water shortages are intensifying in

New Mexico, Texas, Nevada and Utah.

• *Midwest States:* The farm belt of the US faces a lethal combination of drought and dried-up wells. Here, the Ogallala (or High Plains) aquifer located under eight states, the largest single underground body of water in all of North America, which irrigates 3.3 million hectares of farm land, is being drained at a rate fourteen times faster than nature can restore it. Half the Ogallala water is now gone. In metro Chicago, studies now warn that water demands will rise another 30 percent by 2025, thereby requiring a major escalation of bulk water transfers from Lake Michigan. And in 2004 half of Kentucky's 120 counties had water shortages.

• *Southeast States:* These states continue to encounter growing water shortages. The Florida aquifer system, which covers 200,000 square kilometres, is currently being tapped at a rate that is far faster than it can be naturally replenished. Indeed, the water table has dropped so low in Florida that some say seawater is now invading its aquifers. Recently plagued by periodic droughts, a "water war" is developing in the region as Florida, Alabama, and Georgia struggle for access to, and control over, limited water supplies. As the city of Atlanta runs out of drinking water and turns to sources such as the Tennessee River to solve its problems, neighbouring states are vigorously objecting to these kinds of inter-basin transfers.

For decades, water shortages in these and other regions of the US have been circumvented by diverting and transporting water from one basin to another through dams, canals and pipelines. But now, there are few, if any, viable water basins left in the US to serve this purpose. Increasingly, attention has been shifting to Canada and the prospects of bulk water exports from its vast array of lakes and rivers. Here, three giant water diversion and export schemes have been on the drawing boards of engineering firms. First is the North American Water and Power Alliance plan, designed to reverse the flow of northern rivers in British Columbia to run through a series of dams southward into the Rocky Mountain Trench, where the water would then be directed by canals into Washington State, primarily for delivery to the southwestern US states. Second, the Central North American Water Project calls for a series of canals and pumping stations linking Great Bear Lake and Great Slave Lake in the Northwest Territories to Lake Athabasca and Lake Winnipeg and then the Great Lakes for bulk water exports to the US, primarily the midwestern states. Third is the Great Recycling and Northern Development Canal plan, which calls for the damming and rerouting of northern river systems in Quebec in order to bring fresh water through canals down into the Great Lakes, from which it would be flushed into the American southeastern and midwestern states.[99]

Although plans for these huge water export schemes have been around for decades and some of them have been hotly debated in the past, it is quite likely they will be revived in one form or another in response to increasing public anxieties in the US over imminent water shortages. At the same time, there are mounting concerns about the ecological dangers of large-scale extractions from water basins. To date, there is sufficient evidence to support the claim that draining massive amounts of water from lake and river basins disrupts local ecosystems, damages natural habitat, reduces biodiversity and dries up aquifers. During inter-basin transfers, parasites, bacteria, viruses, fish and plants from one body of water are carried into another. Mercury contamination from the flooding required for water diversions eventually accumulates in the tissues of birds and mammals, as it inevitably makes it way up the food chain. As well, the large-scale structures required for the storage of water to be exported will disrupt ecosystems in remote areas such as the Rocky Mountain Trench or James Bay in northern Quebec.

In the foreseeable future, the depletion of fresh water may well become most acute in Alberta, Saskatchewan and the Northwest Territories where the tar sands industry is operating. As we shall discuss in more detail in Chapter 5, the tar sands industry uses a great deal of water in its strip mining and in-situ processes, both for extracting the bitumen and upgrading the bitumen into synthetic crude oil. And water scientists are already warning that renewable water supplies in the tar sands and surrounding prairie regions are in danger of being depleted, thereby raising the spectre of drought conditions. According to studies conducted by Canada's dean of water scientists, Dr. David Schindler at the University of Alberta, the Prairies are already drying up: the South Saskatchewan River has declined 80 percent, the Old Man and Peace Rivers are down 40 percent, while the Athabasca River has dropped by 30 percent.[100] What's more, global warming will no doubt further accelerate these trends: glacier meltdowns will terminate the annual spring runoffs that feed river systems.

Taken together, the convergence of these three planetary forces — peak oil, climate change and the depletion of natural resources such as fresh water — constitutes the triple crisis that both threatens and challenges the future of our high-tech industrial society. As James Howard Kunstler warns in his book *The Long Emergency*, "we are entering an uncharted territory of history," namely, "the long arc of depletion," which will generate "economic, political and social changes on an epochal scale": "Industrial civilization is in big trouble, [and] … people are sleepwalking into a future of hardship and turbulence." With the era of cheap fossil fuels coming to an end and with the onslaught of climate change and natural resource depletion upon us, says Kunstler, there is likely no combination of alternative fuels that can power our economy and society to the degree to which we have become accustomed. The marvels and comforts of modern industrial life

will soon start to disappear, the consumer economy will gradually vanish, and every activity of daily life, from farming to
schooling to retail trade, will be downscaled. Suburban life will
become untenable for most people, the days of easy motoring
and commercial aviation will become a thing of the past, and the
struggle to feed ourselves to survive will intensify.[101]

Today, the tar sands mega-machine is caught in the
crosshairs of this convergence of the triple crisis. More than
any other sector of the Canadian economy today, the tar sands
embody the contradictory and deadly forces at work. How the
peoples of Canada deal with the emerging ecological, economic and social struggles over the development of the tar sands
will, in large measure, determine our destiny as a nation in the
twenty-first century.

CHAPTER 3

Energy Superpower

The challenges posed by the triple crisis have not yet registered more than the smallest blip on Canada's radar screen. Instead of being seen as a microcosm of the convergence of peak oil, climate change and water depletion on the planet, the Alberta tar sands are viewed as signifying the dawn of a new economic age for Canada. If anything, the tar sands industry is considered in some circles to be Canada's answer to the peak oil challenge, the silver bullet that will save industrial civilization from collapse. Indeed, we have a prime minister who believes that development of the tar sands is Canada's launching pad to becoming the world's new energy superpower.

In his first speech to a business audience since becoming prime minister, Stephen Harper declared that Canada was the world's new energy superpower. The speech was delivered July 2006 on the eve of the G-8 Summit of Industrialized Nations, the economic elite of the political world. Harper spoke to the

Canada–United Kingdom Chamber of Commerce in London, before going on to St. Petersburg, Russia, for the Summit.[102] "One of the primary targets for British investors," said Harper, "has been our booming energy sector. They have recognized Canada's emergence as a global energy powerhouse — the emerging 'energy superpower' our government intends to build."

To substantiate his claim, Harper went on to say: "We are currently the fifth largest energy producer in the world. We are ranked third and seventh in global gas and oil production respectively. We generate more hydro-electric power than any other country on earth. And, we are the world's largest supplier of uranium." He then zeroed in on the Alberta tar sands, saying they are "akin to the building of the pyramids or China's Great Wall." With the tar sands, he emphasized, "even now, Canada is the only non-OPEC country with growing oil deliverability." For these reasons, he told his investor audience, Canada is "the most attractive combination of circumstances for energy investment of any place in the world."

While it is not exactly clear what Harper had in mind in promoting Canada as an "energy superpower," he has since repeated this theme several times, including, almost exactly a year later, in a speech delivered at the APEC (Asia Pacific Economic Co-operation) Summit in Sydney, Australia, in August 2007.[103] Nor is it clear from his speeches what economic vision the government has in mind when it says it "intends to build" Canada as an "emerging superpower." What is clear is that the tar sands are the centrepiece of the vision underlying the claim. Hence, it is worth probing a little deeper into the recent history of the development of the tar sands industry to see if there are grounds to substantiate Harper's claim, starting with the corporations involved.

Corporate Players
Stephen Harper's musings about Canada's new-found status as an "energy superpower" were not pulled out of thin air. For well

over a half century, some of the major corporate players in the Alberta oil patch have been working with both federal and provincial governments to build a powerful tar sands industry. The corporate players include the two leading pioneers in tar sands production, Syncrude and Suncor, which have, in many ways, laid the foundations for the industry. Through their initial experiments with technologies for open-pit mining and in-situ production, and through their upgrader facilities, these two companies have led the way in getting the bitumen out of the ground and converting it into synthetic crude. Now Syncrude and Suncor have been joined by a host of other corporate players from the oil industry, including US giants ExxonMobil (via Canadian subsidiary Imperial Oil) and ConocoPhillips, international conglomerates such as Shell Oil and British Petroleum (BP) and a battery of Canadian-based companies such as Canadian Natural Resources Ltd. (CNRL) and Petro-Canada.

In his seminal study of the tar sands, Larry Pratt traces the origins of the industry back to the early decades of the twentieth century.[104] In particular, Pratt shows how Syncrude was able to establish the terms and conditions for the operation of the tar sands industry in Alberta. At the time, Syncrude was a consortium of four US-controlled oil companies: Exxon (Imperial Oil), Gulf Oil, Atlantic Richfield, and Cities Service. Between 1972 and 1975, Syncrude mounted a powerful lobbying machine that influenced decisions at the highest levels of government in Alberta and Canada concerning pricing, taxation, environmental regulation, labour legislation and other policies affecting the development of the tar sands industry.[105] Moreover, the Syncrude consortium managed to convince both the federal and provincial governments to provide the burgeoning tar sands industry with generous subsidies and to underwrite much of the risk associated with tar sands production, in return for little in the way of actual control. In September 1973, then premier Peter Lougheed went on province-wide television to announce that a deal had been struck to launch the giant, billion-dollar Syncrude project after ten years of planning.[106] Fifteen months later, when Atlantic Richfield pulled out of the consortium, another round of intense negotiations began, which resulted in a government rescue plan for Syncrude involving "huge infusions of public cash."[107]

Together, these events laid the foundations for what has become a booming and increasingly profitable tar sands industry. In the thirty-five years since the corporate–government deal was struck to launch Syncrude, numerous oil corporations and corporations in related industries have invested in the tar sands. Today, there are more than sixty-nine tar sands projects under way and many more on the drawing boards for the next ten years. These projects involve either open-pit mining or in-situ production, as well as upgrader facilities. Some of the major corporate players are engaged in all three of these operations. Although several of the corporate players have been Canadian-based companies and consortia, the number of foreign, particularly US, oil corporations investing in the tar sands has been on the rise. In 2007, industry analysts estimated that investment in the tar sands would rise from 30 billion dollars to 125 billion dollars within the next decade.[108] Furthermore, the tar sands industry has considerable room to grow. Besides the lands already leased in the Athabasca, Cold Lake and Peace River fields, more than 65 percent of the lands designated for tar sands operations are still available to be leased for exploration and production.

Today, 10 major corporate players dominate production in the tar sands (see map on page 2). The following is a brief profile of their operations.[109]

• **Syncrude** is a consortium of Canadian Oil Sands, ConocoPhillips, Imperial Oil (ExxonMobil), Petro-Canada, Nexen Oil Sands and a few smaller companies. Its main operations in the tar sands are strip mining and upgrading at their Mildred Lake and Aurora sites. Currently, Syncrude is producing around 407,000 barrels a day (bpd) through both its strip-mining and upgrading operations. By 2015, Syncrude expects to be producing close to 600,000 bpd, most of which will be destined for markets in the US.

- **Suncor**'s involvement in the tar sands dates to the early 1960s when Sun Oil in the US, led by J. Howard Pew, invested a quarter of a billion dollars in the Great Canadian Oil Sands project. In 1979, Suncor Inc. was formed from a merger between Sun Oil and Great Canadian Oil Sands. Almost two decades later, Suncor has become the largest producer of crude oil from the tar sands: it is expected to be generating 344,000 bpd by the end of 2008, and 520,000 bpd in 2012 through both its strip mining at Steepbank and in-situ operations, plus its Tar Island upgrader. By 2015, Suncor's daily output is expected to be 700,000 bpd.

- **Shell Oil,** the international conglomerate, has the controlling interest (60 percent) in the Albian Sands Energy Project, which includes strip-mining and upgrading operations. Shell's partners include two US-owned oil companies, Chevron Canada, a subsidiary of ChevronTexaco, and Marathon Oil. Albian Sands has two strip-mining operations: Muskeg River Mine and Jackpine Mine. From these two mines, Shell intends to raise Albian's daily production from 155,000 to 270,000 bpd by 2010. Albian's total production is expected to reach 570,000 bpd by 2015.

- **ConocoPhillips,** the third largest US petroleum company, has made two major investments in the tar sands so far. The first is a joint venture with the EnCana Corporation, a subsidiary of Enbridge Pipelines of Canada, involving in-situ operations at both Christina Lake and Foster Creek, The second is a joint venture with Total SA of France, called the Surmont Project, which is facilitated through its subsidiary Total E&P Canada. By 2015, these projects are expected to be producing a total of 500,000 bpd.

- **ExxonMobil**, the world's largest oil giant, has established a toehold in the tar sands through its Canadian subsidiary Imperial Oil. In addition to owning 24 percent of Syncrude, Imperial Oil operates an in-situ facility at Tucker Lake and more recently has been developing a strip-mining operation at Kearl Lake. Midway through the next decade, therefore, Imperial expects to be producing over 450,000 barrels a day from its Kearl Mine and Tucker Lake operations. Moreover, Imperial has room to expand much further, given that the company holds leases on 188,000 hectares.

- **Canadian Natural Resources Ltd. (CNRL)**, a Calgary-based oil company, has been positioning itself to be a major player in the tar sands. CNRL's strip-mining operations were launched in 2008 and are forecast to expand rapidly. In 2012, CNRL will add an upgrader facility to its operations and three years later is scheduled to begin in-situ production as well. By 2015, CNRL's strip-mining and in-situ daily production capacity could reach 505,000 bpd, if current plans materialize.

- **Petro-Canada,** to a lesser extent, is also reckoning to become a key player in the tar sands. Besides holding a 12 percent interest in the Syncrude consortium, Petro-Canada possesses controlling interest (55 percent) in the Fort Hills Energy Project. The Fort Hills strip-mining operation is expected to come into production in 2011, and it is projected that by 2015 it will be producing 190,000 bpd. In addition, Petro-Canada has in-situ production facilities at Dover and Mackay River, which are expected to be producing between 70,000 and 80,000 bpd by 2009; it also has in-situ facilities at Chard, Lewis, and Meadow Creek.

• **Husky Energy** is 70 per cent owned by Li Ka-shing, 35 percent directly, and 35 percent through his holding company, Hutchinson Whampoa. Although Husky's in-situ production facility at Sunrise only began operation in 2008, it is expected to be producing 200,000 barrels daily by 2014. Combining this with its in-situ plant at Tucker, which has been operating since 2006, Husky Energy is expected to be producing 230,000 bpd from the tar sands by 2015. Husky also plans to expand the capacity of its upgrader facility at the Saskatchewan–Alberta border city of Lloydminster, which has been operating since 1992.

• **Sinopec**, China's largest refiner and marketer of petroleum products, is also about to launch its strip-mine and upgrader facilities, Northern Lights, in the Alberta tar sands, with its Canadian subsidiary, Synenco Energy. Mining is expected to begin in 2009, and to be producing over 100,000 bpd by 2011. In addition, the Northern Lights upgrader, being constructed alongside the mine, will begin operations in 2010.

• **Total SA** of France is another international player in the tar sands. In addition to its joint venture with ConocoPhillips in the Surmont Project (see above), Total has controlling interest in the Joslyn Mine leases, which cover an area of 221 square kilometres, located north of the Surmont leases. By 2015, Total's Joslyn strip-mine operations are expected to be producing 142,000 bpd. Total is also building upgrader facilities at both Joslyn and Surmount, expected to be in operation by 2010, along with in-situ production facilities.

These are the ten top players in the Alberta tar sands industry to date. Other international oil giants such as British Petroleum (BP), which owns substantial leases in the Cold Lake region of the tar sands, have yet to announce their development plans.

However, BP has made extensive investments in the refining of crude oil from the tar sands in the US.

From these brief profiles and the chart below (Table 3.1), it appears that that the tar sands industry is well on its way to quadrupling its current rate of crude oil production. If all of these plans come to fruition, the tar sands will be yielding close to five million bpd by 2015. Well over half of this production will come from strip mining, which entails the tearing down and ripping up of the boreal forests and digging massive craters into the earth's crust in order to get at the bitumen that lies beneath. Through all of these strip-mining and in-situ production projects, the tar sands industry is gearing up to meet export targets for crude oil to the US.

It should be kept in mind, however, that these figures are targets, planned rather than actual. A number of factors could well come into play to slow down the pace of production, including investment constraints, skilled labour shortages, pipeline construction delays and government approvals. In part, this is why observers generally predict that tar sands production will be around 3 million barrels a day by 2015. Nevertheless, it is important to note that the industry's own production targets outlined above are considerably higher and on schedule to meet growing US demands and market projections.

Before the bitumen can be transported by pipeline to refineries it must "upgraded" in industrial facilities called "upgraders." Using intense heat and pressure, these upgraders take the tar-like bitumen extracted from the sands and "upgrade" it so that it becomes more like crude oil. This process requires a great deal of energy and water. Companies such as Syncrude and Suncor have their own upgraders on site near their strip-mining and in-situ

*Right: *Data compiled from National Energy Board and industry reports by the Polaris Institute.*

Table 3.1

Tar Sands Projects: Barrels of Oil Produced Daily in 2008 and Projected for 2015*

Company	Strip Mine 2008	Strip Mine 2015	In-situ 2008	In-situ 2015	Total 2008	Total 2015
Syncrude	407,000	593,000			407,000	593,000
Suncor	276,000	324,000	68,000	376,000	344,000	700,000
Albian Sands Energy Project Shell Canada, Chevron Canada and Marathon	155,000	570,000			155,000	570,000
ConocoPhillips, EnCana Joint Venture			70,000	400,000	70,000	400,000
ConocoPhillips, Total SA Joint Venture				100,000		100,000
Total SA		100,000	12,000	42,000	12,000	142,000
Imperial Oil (ExxonMobil)		200,000	140,000	170,000	140,000	370,000
Canadian Natural Resources		415,000		90,000		505,000
Fort Hills Energy Project Petro-Canada, UTS Energy Corporation and Teck Cominco		190,000				190,000
Husky Energy			30,000	230,000	30,000	230,000
Sinopec		100,000				100,000
Other		262,000	48,500	933,500	48,500	1,195,500
Total	838,000	2,754,000	368,500	2,241,500	1,206,500	4,995,500

Total projected output from all major tar sands operations by 2020   5,255,500 bpd

operations. However, much of the increase in the production of tar sands crude outlined above will take place in facilities planned for construction northeast of Edmonton. As many as nine upgrading facilities are to be built along the North Saskatchewan River in the counties of Strathcona, Sturgeon and Lamont.[110] Now called "Upgrader Alley," this is expected to be the core of Alberta's industrial heartland. Here, one upgrader is already in operation, two more are under construction, five more are going through the application process, and land has been acquired for another. Each upgrader requires hundreds of hectares of land, most of which is currently used for agriculture in this region. Upgrader Alley will be fortified with new infrastructure such as electrical transmission lines and railway lines.

To transport this rapidly expanding volume of tar sands crude

The Opti-Nexen upgrader and related facilities at Long Lake, south of Fort McMurray, Alberta. Courtesy of David Dodge, Canadian Parks and Wilderness Society.

to US markets requires major increases in pipeline capacity. At present, there are twenty-two pipelines linking oil production in Canada (mainly Alberta) to the five major petroleum market regions in the US (see Figure 3.1).[111] However, new pipeline systems will be needed to transport the planned five-fold increase in tar sands production to US markets. One major pipeline project is the Alberta Clipper being constructed by Enbridge Pipelines Ltd. The Alberta Clipper is a 1,607-kilometre pipeline designed to ship tar sands crude and bitumen from Alberta to refineries in Wisconsin, starting in mid-2010. When completed, the Alberta Clipper will initially move 450,000 bpd and eventually up to 800,000 bpd to US refineries and markets. Enbridge is also involved in the construction of a southern access pipeline that will transport tar sands crude further south to refineries and markets in Texas and the Gulf Coast.

Additional pipelines are in the works, some already being constructed, others in the planning stage, to link the Alberta tar sands to the growing demand for crude oil in the US. The Trans Mountain Pipeline, for example, from Edmonton through British Columbia to Washington State, is due to be completed in November 2008. According to US officials, this pipeline is urgently needed to supply Alberta crude to refineries in Washington State that serve US military operations on the Pacific Coast. Then there is the Keystone Pipeline, which is specifically designed to carry upgraded tar sands crude to refineries in southern Illinois. The Keystone will be 3,456 kilometres long: the Canadian portion will be built by TransCanada Pipeline. The National Energy Board has approved the Keystone project and it has received the green light from US regulatory bodies. When the Keystone Pipeline is completed, it will be able to transport 590,000 bpd to refineries and markets in the Midwest.[112]

Most of the heavy crude from the tar sands is transported in upgraded form to the US where it is refined and sold for all sorts of commercial uses. Companies such as British Petroleum and ConocoPhillips, which are already committed to the development of the Alberta tar sands, have also made major investments in developing special refining capacities in the US for this heavy crude. BP, for example, is spending 3 billion dollars to reconfigure its refinery in Whiting, Indiana, so that it can refine the heavy crude that comes from the tar sands. According to BP's website, the Whiting refinery on an average day "produces enough products to fuel 430,000 automobiles; more than 10,000 farm tractors; 22,000 semi-trucks; 2000 commercial jet liners"; and to fill 350,000 propane cylinders and produce 8 percent of all the asphalt used in road construction across the US.[113] By 2011, up to 90 percent of the crude for BP's Whiting refinery will come from the tar sands. Similarly, ConocoPhillips is investing more than 1 billion dollars to expand the crude and coking capacity of its Wood River refinery in Illinois so that it can refine more heavy crude from the Alberta tar sands. In June 2008, the Environment Integrity Project (EIP) in Washington reported that plans were being made to expand the capacities of many more existing refineries in the US to process tar sands crude.[114] The report identifies eleven states where oil refinery capacities are to be expanded to process over 1.9 million bpd. According to EIP, these expansions are the equivalent of sixteen new refineries.

Meanwhile, in July 2008 Shell announced that it has scrapped its multi-billion dollar plans to build a new refinery near Sarnia, Ontario, for tar sands crude. Instead of transporting the bitumen from the tar sands down south to the US to be upgraded, Shell's proposed refinery in Sarnia was initially viewed by governments as an important step towards value-added production and job creation in Canada. Now Shell claims that it would be more cost effective for the company to expand its existing refineries in the US and is looking at retooling its refineries in Martinez, California, and Deer Park, Texas, to take bitumen. In 2007, Shell did file

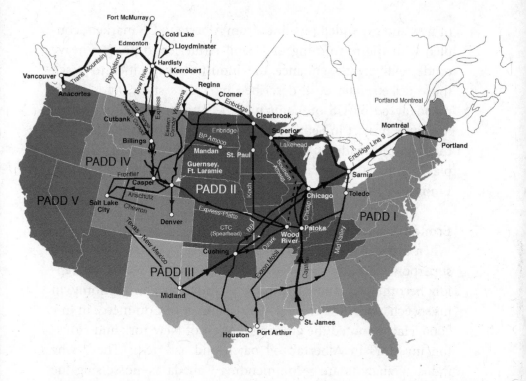

Figure 3.1. Major Canadian and US Crude Oil Pipelines and Markets. Source: National Energy Board.

plans with Alberta regulators to build the largest upgrader plant in the tar sands to date. The decision whether to build the proposed Shell upgrader plant, which would eventually have the capacity to handle 400,000 bpd of bitumen, will be made in 2010.

In effect, the infrastructure has been laid for the tar sands industry to extract, produce, deliver and refine crude oil to fuel the US. If the production plans outlined above by the major corporate players involved in strip mining and in-situ production of tar sands crude for export come to fruition, as scheduled, then they will be well on their way to meeting the five-fold increase targets that were set at the Houston summit in January 2006 (described later in Chapter 4). As well, the plans for the construction of new and expanded pipelines from Alberta to US markets, coupled with the retrofitting of US oil refineries to process heavy crude, will greatly enhance the industry's capacities to reach those targets. The deal, described by Pratt's study, and largely manoeuvred by US oil companies, led to the launching of the Syncrude project in September 1973 and laid the foundation for the tar sands industry today. By putting this infrastructure in place, Canada's role as the energy satellite of the American Empire will be consolidated.

Economic Backbone

When Stephen Harper declared Canada the world's new energy superpower, he also signalled that the oil and gas sector was rapidly becoming the new backbone of the Canadian economy. In his speech before the Canada–UK Chamber of Commerce in July 2006, Harper noted the growing interest of New York and Houston investors in Alberta's oil patch and asserted: "That's why industry analysts are recommending Canada as 'possessing the most attractive combination of circumstances for energy investment of any place in the world.'" With world oil prices beginning to soar and the Canadian loonie taking flight, global currency markets were already signalling the emergence of Canada's petro-dollar. Canada's economy is undergoing significant changes and the Alberta tar sands are playing a central role.

For Harper the Alberta oil patch is part of a larger vision. When speaking about Canada's energy superpower status, Harper stresses this country's role as the biggest generator of hydroelectric power and the largest supplier of uranium for nuclear energy plants. But Harper's vision also contains a northern component, namely, the development of the Arctic through the opening up of the Northwest Passage. Reminiscent of John Diefenbaker's Conservative government's "northern vision" a half century ago, Harper began highlighting a northern development vision of his own for the country's future in the summer of 2007. While Diefenbaker's plan was to open up a mid-Canada corridor, Harper's strategy is focused further north, on the Northwest Passage. Promoting Canada as "The True North Strong and Free" renewed could boost Harper's popularity in the country, argues author and columnist Lawrence Martin.[115] It could also begin to answer the question author Michael Byers says is on people's minds these days: "What is Canada for?"[116]

As former Prime Minister Paul Martin's government did, the Harper government has made Canada's economic and political sovereignty over the continental shelf in the Arctic a top priority. With global warming, the normally icebound passage through the Arctic Ocean could soon be open to commercial ship traffic in the summer months, thereby opening up the east–west trade route at the top of the world that has eluded traders and industrialists for centuries. Canada has until 2013 to submit documentation of its sovereign claim to the outer limits of its continental shelf to the United Nations Commission on the Limits of the Continental Shelf. Three federal departments (Natural Resources, Fisheries and Oceans, and Foreign Affairs) are working together to prepare Canada's case. The government's objective is to complete the mapping of the continental shelf and to prepare a submission to the UN that establishes, as one of Harper's ministers put it, "Canada's ability to explore and exploit its natural resources over the 1.75 million square kilometres of its extended continental shelf — an area about the size of the Prairie provinces."[117]

The Harper government is also well aware that the Northwest Territories, stretching from the Alberta border to the Beaufort Sea, is endowed with minerals ranging from diamonds, platinum and uranium to gold, silver, lead and zinc, as well as oil and natural gas. Hence, this northern vision includes the construction of north–south railway lines and deep sea ports along the Arctic coast for cheaper transportation (including LNG tankers for natural gas) of mineral resources south, and equipment and personnel north. In its 2008 budget, the Harper government began to put its "Vision for the North" into play by announcing mineral exploration tax credits, sovereignty incentives to attract workers, a commercial harbour in Nunavut, docking and refilling facilities along Canada's northern coastline and additional Arctic patrol ships, as part of a more comprehensive northern strategy. By making these investments now, the Harper government wants to show Canada is serious about its claims to Arctic sovereignty.[118]

The centrepiece of this economic vision appears to be an energy mega-corridor, centred in Alberta, linking the Arctic in the north to the US in the south through networks of pipelines and other industrial infrastructure. For the time being at least, the Alberta oil boom is the engine that is driving this economic vision. But, the real driver behind the Alberta oil boom and the energy corridor is the tar sands. It is the Alberta tar sands that makes Canada, in Harper's words, "the only non-OPEC country with growing oil deliverability." If the Mackenzie Gas Project (MGP) is constructed — which is primarily designed to transport natural gas from the Mackenzie Delta and the Beaufort Sea down south, to fuel the tar sands production processes as well as to service southern markets — it will be a strategically important part of the energy mega-corridor. Moreover, these energy corridor plans were given an added boost in July 2008 when the US Geological Survey issued a report saying that up to 90 billion barrels of "technically recoverable" oil may be available in the Arctic Circle, much of which lie in waters currently claimed by Canada, Russia, Norway, Denmark and the United States. Meanwhile, the development of the tar sands is being accelerated by skyrocketing oil and other commodity prices, which, in turn, are largely responsible for the spike in the value of the Canadian dollar, all of which is having a profound impact on the Canadian economy.

One might ask, of course, how much of this series of developments is really the result of an economic vision. With the exception of protecting Canada's sovereignty over the Arctic, and the construction of the MGP and mining and petroleum operations in the Northwest Territories, Canada does not have much control over what happens in this energy mega-corridor. Under the Canada Lands Act, enacted as part of the 1980 National Energy Program, the federal government controls the resources of the Northwest Territories. But the Harper government does not have the necessary tools in place to shape the development of the North, such as a state-owned oil company to control production. Instead, the vision of an energy mega-corridor is largely driven by US market demands and the US petroleum industry, solidly backed, for the time being at least, by the energy strategy of the Bush–Cheney administration in Washington. As a result, Harper's pronouncements may amount to little more than a rationalization of the plans that are already being set in motion by these driving forces. Nevertheless, one of the federal levers that the Harper government has continued to use to promote this strategy is a program of subsidies to the oil and gas industry in the tar sands.

The Harper government, like the Martin and Chrétien governments before it, strongly supports the tar sands industry through federal subsidies. Between 1996 and 2002, corporations operating in the tar sands received almost 1.2 billion dollars in federal subsidies: 625 million dollars in tax writeoffs, 507 million dollars through the Syncrude Remission Order, and 60 million dollars for research and development.[119] Moreover, under the Harper government, tar sands companies are treated more favourably than conventional oil companies. Since the 2006 budget, tax writeoffs for property and pre-development costs are more generous for tar sands projects, both strip-mining and in-situ, than they are for conventional oil operations. For example, the most lucrative of these federal subsidies, the Accelerated Capital Cost Allowance (ACCA), allows tar sands corporations to write off their costs for new projects and expansions from federal and provincial taxes until all of their capital costs are paid off. By comparison, conventional petroleum companies are permitted to write off only 25 percent of these costs per year.[120] According to Finance Minister Jim Flaherty, the ACCA for tar sands projects will be worth 300 million dollars a year on average over the period 2007 to 2011. Over this five-year period, therefore, the tar sands industry will receive a federal subsidy totalling 1.5 billion dollars from this tax measure alone.[121]

At the same time, tar sands companies are granted lucrative provincial subsidies in the form of low royalty rates. In Alberta, the Klein government required that companies operating in the tar sands only pay a royalty fee of 1 percent on their gross revenues until all their construction costs are paid off, which effectively amounts to a tax holiday. After that, they would pay a 25 percent tax on their net earnings. In contrast, oil-producing countries such as Russia, Bolivia and Ecuador collect 90 percent or more of the windfall profits that the companies reap once the price of oil goes above a certain level (e.g., in Russia, 25 dollars a barrel).[122] Rejecting such windfall tax schemes, the new Stelmach government announced a much more modest royalty regime. If oil prices jump from 55 to 120 dollars a barrel, then the royalty rate would increase from 1 to 9 percent and, similarly, the tax rate would rise from 25 percent to 40 percent. If, for example, the world price for oil reaches 120 dollars a barrel (as it did in April 2008), then the Alberta tax rate on the net earnings of the tar sands industry would be 40 percent, which is still very generous compared to most other regimes.

With these kinds of government subsidies, it is clear that Canada and Alberta have made supporting the tar sands industry a top priority. Although the 2007 budget of the Harper government included the phasing out of the Accelerated Capital Cost Allowance, which accounts for some 77 percent of the federal subsidy to the tar sands industry, this will happen gradually. Those companies that began their projects or expansions before March 19, 2007, will be able to claim the 100 percent writeoff, while for those companies that began their projects after this date the tax subsidy will be phased out between 2011 and 2015.[123] The underlying objective of the 2007 budget measure was to give the appearance of repealing the special ACCA tax break for tar sands companies without actually doing so until after 2015. The very slow phase-out has the effect of providing incentives worth 1.5 billion dollars over the period 2007 to 2011 to stimulate investment and speed up the starting of new projects and expansions in the tar sands. All this is being done in order to meet projected demands for crude oil in the US. In other words, to encourage more companies to get out their gigantic shovels now and start ripping out the bitumen before the federal subsidies run out. Meanwhile, Ottawa's entire system of tax subsidies to the oil industry has come in for some strong criticism from the Organization for Economic Co-operation and Development (OECD).[124]

As suggested above, these moves appear to be aimed at realizing a larger economic vision. Subsidizing the tar sands industry is not simply a matter of picking "winners" and "losers" among corporations and industries in a highly competitive climate. Instead, there appears to be a deliberate strategy to restructure the Canadian economy with a renewed emphasis on the resource sector. As economist Jim Stanford has noted, throughout the second half of the twentieth century, federal and provincial governments were proactively moving the Canadian economy away from being primarily a resource hinterland.[125] Measures to facilitate this included encouraging and supporting the development of processing and

secondary manufacturing of resources; more sophisticated supply industries to feed into resource production activities; other
high-tech manufacturing industries (e.g., the aerospace industry)
and the expansion of industries providing exportable services. By
the mid-1990s, says Stanford, Canada had become "a global manufacturing powerhouse." But this industrial strategy petered out
around the dawn of the twenty-first century and steps were taken
to restructure the Canadian economy around natural
resources.[126]

Over the past two decades, governments have adopted a much
more *laissez-faire* approach to the Canadian economy. National
economic development has been largely market-driven, led by
the free trade regimes, notably, the Canada–US FTA and NAFTA.
As a result, Canada's economy has been undergoing a "historic
structural shift" from a more diversified industrial economy to a
resource-based economy (especially energy resources). As we
have seen, the proportional sharing rule of NAFTA explicitly
assigns Canada the role of energy storehouse for the US economy. Canada's economic destiny, therefore, is increasingly being
shaped not only by global market forces and private investment
decisions, but by US energy security strategies. On the one hand,
this restructuring of Canada into an energy and other resource
supplier to US and global markets has generated an oil and
resource boom that has contributed to relatively healthy overall
growth and job creation. On the other hand, there are many negative consequences as well.

This resource-led restructuring of Canada's economy has been
spurred on by the rise in commodity prices in world markets.
Between 2007 and 2008, the price of oil virtually doubled. The
skyrocketing of oil prices to well above the 140-dollars-per-
barrel mark has certainly made tar sands oil production a highly
profitable endeavour, especially given a hungry export market in
the US that wants to break its dependency on politically unstable

petroleum sources elsewhere in the world. But these factors have also combined to create a dramatic rise in the value of the Canadian dollar on global currency markets. The loonie has been moving upwards against the US dollar since late 2002. Currency traders tend to value the Canadian dollar in relation to global commodity prices. When mineral prices, especially the price of oil, rise or fall, so too does the value of the Canadian dollar. Currency traders, in other words, now view the Canadian dollar as "petro-currency."[127] And a stronger Canadian dollar on global currency markets has meant imported products are cheaper, but it has also had a devastating impact on other Canadian industries that depend on exports, especially to the US.

One of the major casualties has been our manufacturing sector. During the post-war period, notes Stanford, Canada gradually built a more diversified and productive industrial base, using mechanisms such as the Auto Pact and Technology Canada Partnerships to spawn high-value industries ranging from aerospace and specialty vehicles to telecommunications equipment and high-tech machinery. In 1999, Canada's automotive industry was ranked as the fourth largest assembler of motor vehicles in the world, and, for the first time in our history, the country exported more manufactured goods than we imported. Since 2002, however, when Canada's employment in manufacturing peaked, this country has lost over 400,000 manufacturing jobs. By the end of 2007, the number of jobs in the manufacturing sector as a proportion of Canada's total employment had dropped by one-quarter, the lowest in the post-war period.[128] At the same time, the country's manufacturing trade deficit ballooned to the point where it exceeded 30 billion dollars in 2007, including the largest automotive trade deficit in our history.

Although service has been a rapidly growing sector in the Canadian economy, there are good reasons for maintaining a strong and vibrant manufacturing sector. Take, for example, global trade. While only 10 percent of the service sector (mostly private companies) in this country is involved in cross-border trade, manufactured products account for 75 percent of Canada's total merchandise trade. In contrast to the generally lower paying jobs in the service sector, manufacturing jobs are for the most part well paid and of higher quality. Most economists recognize that the manufacturing sector has, on average, higher rates of productivity than most other sectors of the economy which, in turn, allows employers to pay wages that are often 25 percent higher than the average wage. What's more, manufacturing industries generally invest a much higher percentage of revenues in research and development (R&D) than either the service or resources sector. As Stanford points out, well over half of all non-government investment in R&D in Canada comes from the manufacturing sector.[129]

However, turning Canada into an energy and resource supplier, while allowing the manufacturing sector to deteriorate, will not improve our balance of trade. In 2006, despite the oil boom under-way, petroleum exports accounted for only 12.6 percent of Canada's total exports. That year, oil and gas exports were worth 66 billion dollars, while the automotive sector recorded 75 billion dollars worth of exports.[130] Moreover, notes Stanford, Canada's oil exports are offset by oil imports. Close to 60 percent of Canada's total oil needs are met by imports to the eastern part of the country: Quebec, parts of Ontario and the Atlantic provinces. When these oil imports are factored into the balance of trade equation, Canada's *net* petroleum exports for 2006 were 4.6 billion dollars, or around 1 percent of the country's GDP. In addition, a flood of lower cost imports, plus the rising value of the loonie, has more than cancelled out the growing value of our energy and resource exports. Since 2000, our merchandise trade balance has deteriorated by 50 percent.

But even if petroleum exports were to account for a larger share of Canada's total exports, this would not necessarily strengthen the economic backbone of the country. Without value-added production, a resource-based economy cannot be sustained in the long run, especially one based on a non-renewable resource such as oil. Instead of refining the raw bitumen from the tar sands in Canada for the manufacture of petroleum-based products before export, most of the bitumen is simply transported by pipeline to US refineries. The Keystone Pipeline, for example, is specifically designed for this purpose. Nor has the Alberta government developed much in the way of an industrial strategy to add value through the development of other petroleum-based manufacturing of, for example, petrochemicals, fertilizers and paints. Unions such as the Communications, Energy and Paperworkers Union of Canada (CEP) and the Alberta Federation of Labour have consistently advocated that a comprehensive industrial strategy along these lines be developed to accompany the oil and gas industry in Alberta. But, for the most part, their proposals have fallen on deaf ears.

It is also worth noting that increased foreign investment and takeovers in the resource sector do not necessarily translate into a more productive economy. In 2007, foreign takeovers of resource companies amounted to over 100 billion dollars.[131] But Investment Canada has very lax regulations governing foreign takeovers of Canadian oil companies, including the takeover of licences to exploit particular petroleum reserves. As well, the transfer of ownership from Canadian to foreign hands of petroleum assets and facilities does not, in and of itself, enhance our economic growth. In fact, it can be a drain on the economy, transferring money to foreign-based shareholders. Given oil prices of 120 dollars or more, producers could earn profits of up to 95 dollars on each barrel of oil coming from existing tar sands plants. Assuming there are 200 billion barrels of recoverable oil, the potential undiscounted profit from this resource could amount to between 17 and 18 trillion dollars, at current prices. Not only has the Alberta tar sands become a lucrative investment, but with increased ownership and control by US and other foreign-owned petroleum companies, most of the profits are destined to leave the country.[132]

In many ways, the 2008 survey of Canada's economy by the OECD confirms these observations.[133]

In its surprisingly critical assessment of the energy sector and the tar sands, the OECD concludes that the Alberta "oil boom" is entirely the result of rising oil prices and that real productivity in Canada's energy sector lagged behind the economy as a whole between 2003 and 2007. Since significant new investments in tar sands production have yet to generate major increases in output, productivity continues to sag. By 2020, says the OECD Report, the tar sands industry will have a very limited impact on the Canadian economy: a 1.1 percent increase in Canada's GDP. The main reason, says the OECD, is that tar sands production could be severely constrained by rising natural gas prices, limited water availability, carbon reduction requirements and other development costs in the future. The economic benefits of the construction boom will be short-lived as well. Investments for construction are heavily concentrated for a short time and once construction of the new production plants and upgraders has peaked, many workers will be let go. In the meantime, the report notes, many of the long-term, high-quality jobs in Alberta's manufacturing industries are being pushed out of the province.[134]

Yet, the tar sands industry itself involves a great deal of technological innovation that could, if further developed, strengthen Canada's role as an energy superpower. Take, for example, the scientific research involved in in-situ production or the upgrading of bitumen into light crude, or the massive machinery used in strip mining, such as the 797B Caterpillar trucks (which stand

one-and-a-half stories tall) and the 495HF Bucyrus electric shovels, or the system of upgraders and pipelines that makes it
possible for the bitumen to be transported to refineries more
than halfway across the continent, or the elaborate communications technology required at almost every stage of tar sands
production. In short, the tar sands industry has been a pioneer in
the development of techniques to recover hard-to-get-at hydrocarbon deposits. In other regions of the world, notably
Venezuela, there are similar tar sands deposits where these new
technologies could be utilized. Surprisingly, Harper's economic
vision does not appear to include plans to develop this technology for export. Indeed, the Harper government has done little or
nothing to promote research and development along these lines.

In the meantime, the Alberta oil boom, spearheaded by the
massive tar sands development, is causing a historic shift in economic power and political power. Since before Confederation,
the centre of economic and political power in Canada resided
mainly in the industrial and agricultural heartland of Ontario
and Quebec. With the tar sands boom on the one hand and a
declining manufacturing base on the other, the country's economic and political power base is shifting from east to west. The
more the Canadian economy becomes resource-based again, the
more likely we are to recognize Calgary as its epicentre, rather
than Toronto or Montreal. Similarly, though to a lesser extent,
petroleum developments in Hibernia (Newfoundland) and Sable
Island (Nova Scotia) have strengthened the economic power base
of Atlantic Canada. If the manufacturing sector in Ontario and
Quebec continues to decline, then we can expect these transitions
of economic power to further accelerate. And with economic
transitions come shifts in political power, as illustrated by the rise
of the Stephen Harper Conservatives from their Alberta base to
form the government in Ottawa.

All of this can be seen as part of the historical development of the Canadian federation. A more equitable redistribution of economic and political power would, of course, be an important step forward in the development of Canada's economy and its federation. But the Alberta oil boom is not the model to follow. As we shall see in Chapter 6, redistribution of the bounty generated by the Alberta boom has been trickle-up, not trickle-down, with the lion's share of the wealth going to the oil corporations, not the majority of people in Alberta, let alone the rest of the country. Meanwhile, the majority of Canada's population still resides in Ontario and Quebec, which are rapidly losing manufacturing jobs. If these shifts in economic power continue, tensions and conflicts within the Canadian federation are likely to increase. Recent predictions that Ontario may become a have-not province, and the complaints that followed, exemplify these tensions.

Harper's Firewall

Meanwhile, the Harper government has taken measures to protect the Alberta tar sands as the centrepiece of its claim that Canada is the new, emerging energy superpower. To a large extent, these measures have been designed to protect the development of the tar sands from being slowed by interventions from Ottawa. In other words, the Harper government has taken preventive steps to ensure that the experience of the National Energy Program of the 1980s is not repeated. To do so, the Harper team has been quietly building a "firewall" around the tar sands, in collaboration with the oil companies and the Alberta government.

Surprisingly, little attention has been paid to Stephen Harper's own connections with the Alberta oil patch since he became prime minister. Those connections are close, not only with key players in the Alberta oil patch, but also with the tar sands industry.[135] Take, for example, Gwyn Morgan, the former chief executive officer of the EnCana Corporation, a subsidiary of

Enbridge Pipelines Co., which is engaged with the US oil giant ConocoPhillips in a major joint venture in the tar sands. Morgan has long been an insider in the Harper team and an advisor to the prime minister. After leaving his post as the CEO of EnCana, Morgan operated, in 2005 and 2006, as a registered lobbyist for his former company. Known for his vigorous opposition to the Kyoto Accord, Morgan was once described by the *Globe and Mail* as "the self-appointed head of the Anti-Kyoto lynch mob." During the 2003 election campaign, the last before the new election finance laws were enacted, EnCana made the largest financial contribution to the new Conservative Party of any oil company.[136]

Indeed, a closer look at the Harper team reveals a well-oiled revolving door between the Conservative government in Ottawa and lobbying firms for the tar sands industry. Another example is Ken Boessenkool, a long-time close associate of Stephen Harper's and a fellow economist. Together, they have worked with the Reform, Canadian Alliance and the new Conservative parties. When Stockwell Day was the Treasurer of the Alberta government under Ralph Klein, Boessenkool worked for him as a policy advisor. Between 2002 and 2004, Boessenkool was senior policy advisor to Harper, then Leader of the Opposition in the House of Commons. In 2004, Boessenkool left this position to work for the prestigious and powerful Ottawa lobby firm Hill & Knowlton. There he continues to this day to lobby the Harper government on behalf of the oil industry, including companies operating in the tar sands. His major clients currently include Suncor and Enbridge, and, until 2004, ConocoPhillips Canada.

In early 2001, Boessenkool was one of five people who worked with Stephen Harper to craft his now infamous "firewall" letter to Ralph Klein.[137] At the time, Harper was president of the right-wing National Citizens Coalition and actively promoting "a stronger and more autonomous Alberta." The letter outlined a set of proposals labelled an "Alberta Agenda." Noting that an economic slowdown or even recession threatened North America, the letter warned that "the [Liberal] government in Ottawa will be tempted to take advantage of Alberta's prosperity, to redistribute income from Alberta to residents of other provinces in order to keep itself in Ottawa." The letter went on to say: "It is imperative to take the initiative, *to build firewalls around Alberta*, to limit the extent to which an aggressive and hostile federal government can encroach upon legitimate provincial jurisdiction [emphasis added]." The text was published as an "Open Letter to Ralph Klein" in the *National Post* in late January 2001, and provoked a major public controversy.

The controversy centred around the notion of building "firewalls" around Alberta, which was seen as an aggressively defensive position against the central government. Harper's philosophical soulmate Tom Flanagan, who joined him in heralding the economic theories of Frederich Hayek in support of free markets without government intervention, was a co-signatory of the letter.[138] As Flanagan explained it in his book *Harper's Team*, the intent of the letter was to challenge Klein to take a stronger stance with Ottawa by advocating an Alberta Agenda, much like Quebec had done in the past with its slogan "*maîtres chez nous.*"[139] But the proposals in the letter became known as the "Firewall" rather than the "Alberta Agenda." "Some commentators," said Flanagan, "(most notably Preston Manning) preferred to emphasize the aggressive connotations of 'fire' and the defensive connotations of 'wall' in order to dismiss the whole idea." In response, Flanagan argues that the term "firewall ... is the perfect metaphor to express what we were trying to do," whether one understands it as it is used in the building trades or in the information technology business.[140]

Today, the term "firewall" may also be the perfect metaphor to describe what the Harper government has been doing to protect the Alberta tar sands. After all, the continuation and gradual phasing out of the lucrative subsidies regime and the schedule for regulatory measures on greenhouse gas emissions instituted by the Harper government were, in effect, "firewalls" designed to protect the tar sands industry and promote its rapid expansion. Although the federal subsidy scheme for the tar sands began under the Chrétien government and continued under the Martin government, the Harper team's new subsidies regime provides oil companies with a protective shield of economic and political stability until 2015. As we saw in the last section, it was designed to ensure that new projects be developed and existing ones expanded in order to meet increased production and export targets for the US. And, as we shall see in Chapter 5, the Harper government's 2007 regulations and its schedule for the reduction of greenhouse gas emissions provides a protective shield for the tar sands industry until the end of 2011, while all other industries in Canada must comply immediately. Furthermore, their 2008 regulations on carbon capture and storage do not come fully into effect for the tar sands industry until 2018. Indeed, the Harper government has virtually granted the tar sands industry a free ride when it comes to reducing greenhouse gas emissions. If anything, says a *Toronto Star* editorial ("Emissions Rules Not So Tough," March 12, 2008), the Harper government's 2007–08 environmental regulations could have the perverse effect of accelerating production of tar sands crude to beat the deadline.

These and other firewall protections for the tar sands industry were facilitated to a great extent by the revolving door between the Harper team and the oil corporations and their lobby machines in Ottawa. Geoff Norquay, who frequently appears as a Harper insider on political talk shows, also works for the Earnscliffe Strategy Group, an influential Ottawa lobbying firm. In 2004–05, when Harper was opposition leader, Norquay was his director of communications and a key player in the party's campaign during the 2004 general election. At Earnscliffe, he represented Suncor between 2006 and 2008, and Shell Canada in 2006. Working side by side with Norquay at Earnscliffe is Yaroslav Baran, who was a senior communications strategist for the Conservatives during both the 2004 and 2006 election campaigns. From April 2002 through August 2005, Baran held various communications and public affairs posts in the leader of the opposition's office. In 2006, he worked for the lobby firm Tactix Government Consulting where his clients included Shell Canada and Enbridge.

In short, both Norquay and Baran have been well placed to ensure that the Harper government provide firewall protections and incentives for the tar sands industry. When the next federal election is called, Baran is expected to be heading up communications in the Conservative war room. In January 2008, Norquay told the *Ottawa Citizen* that he makes use of "talking points" prepared by the prime minister's office (PMO) before speaking on radio or television.[141] Norquay also expects to participate in the daily eight a.m. communication strategy meetings of the Conservative Party during the next federal election campaign. Given his public profile as a political pundit and his clientele at Earnscliffe, Norquay has become known in the oil patch as the go-to guy for tar sands companies looking for a lobbyist with a direct line to the PMO.

Yet Norquay and Baran are certainly not the only Harper insiders who are active lobbyists for the Alberta oil patch. Take, for example, Kristen Anderson, who held positions as public affairs officer and outreach officer in the leader of the opposition's office between July 2003 and July 2005.[142] During this three-year period, she played key roles in communications, messaging and event planning, and she worked in the Conservative Party war room during the 2004 election campaign. In 2006, Anderson left to go to work for another Ottawa lobby firm, Global Public Affairs, where she now represents some of the major corporate players in the Alberta tar sands, including Imperial Oil, ConocoPhillips Canada, Chevron Canada, Petro-Canada, Synenco and Teck Cominco. In 2006, Anderson also represented Syncrude, the longest standing player in the tar sands, as well as BP.

Because of this revolving door through which Harper insiders move back and forth between Parliament and lobbying firms for the oil industry, the Conservative government has been able to both maintain and enhance firewall protections for the tar sands industry, particularly when it comes to the crackdown expected on greenhouse gas emissions. Given the rising tide of public opinion calling for effective government action on global warming, there are sufficient grounds to justify direct intervention by Ottawa to control development of the tar sands. This, of course, was the great fear in the Alberta oil patch at the time the Harper team took over the reins of government in Ottawa in January 2006. By this time, the anti-Kyoto sentiment amongst oil company executives had been cranked up once again by insiders such as Gwyn Morgan. Under Harper's first environment minister, Rona Ambrose, the government faltered in its attempts to come up with a plan of action on greenhouse gas emissions that was satisfactory to the favoured tar sands industry. But after Ambrose was replaced by John Baird, a more acceptable plan was crafted. Baird has proven to be effective in continuing to provide the tar sands industry with a protective shield.

Given skyrocketing oil prices, coupled with these kinds of firewall protections, the Alberta tar sands industry is almost guaranteed that the boom will continue, even in the throes of an economic recession. After all, the US needs access to a secure supply of oil and desperately wants to break its dependence on politically unstable supply chains in the Middle East and elsewhere. For Harper, this combination of circumstances further reinforces his claim that Canada is an emerging energy superpower. A strong, flourishing and well-protected tar sands industry that can provide an increasing oil supply, coupled with a nearby, insatiable and guaranteed market, seems to provide grounds for Harper's claim. But are these grounds really sufficient for Harper to claim energy superpower status for Canada? *Globe and Mail* reporter Shawn McCarthy sees Canada as being more of an "energy superstore" than an energy superpower, while author Linda McQuaig describes Canada as an "energy pussycat."[143]

In a feature article for *Alberta Oil*, Sebastian Gault argued that applying the label of "energy superpower" to Canada may be more a sign of "delusions of grandeur" than anything else. Annette Hester, a senior associate at the Center for Strategic and International Studies in Washington, goes further, to suggest that the Harper government does not know what it means to be an energy superpower. In a provocative paper published by the Canadian Defence and Foreign Affairs Institute in Calgary, Hester outlines several standard measures for assessing whether a country is an energy superpower.[144] They include abundant supplies of oil and natural gas and sufficient control to set prices, either through the market or by withholding supplies; a government able to leverage its energy resources to reach political objectives; and the ability to utilize its energy resources strategically to force other countries or companies to do what they would not otherwise do. Measured against these criteria, concludes Hester, Canada is not an energy superpower.

Although, thanks to the tar sands, Canada possesses an abundance of crude oil and, unlike many countries, can produce increasing volumes for the foreseeable future, says Hester, our natural gas output is declining and the production of crude from bitumen is proving to be not only a technical challenge but environmentally costly.[145] In Canada, the industry is owned exclusively by the private sector, and therefore it is the market and not the state that sets prices. In fact, Canada is a "price taker, not a "price setter," notes Hester.[146] Moreover, under the Canadian

Constitution, it is Alberta, not Ottawa, that owns the resource, thereby limiting the amount of control the federal government can exercise. But even if the Harper government wanted to use our energy as a bargaining chip, it could not withhold exports of oil to the US as a show of strength without breaking the rules in NAFTA. Finally, says Hester, most people in Canada do not want to impose our will on other countries, nor do we have the sense of superiority or the go-it-alone-if-necessary attitudes that are characteristic of superpowers.

However, the Harper government proposes portraying Canada as a different kind of superpower. Instead of using our energy resources as military strength to get our way, softer tactics would be used, say officials at Foreign Affairs. In other words, Canada would be "the kinder, gentler energy superpower." As Brian McKenna and David Ebner wrote in the *Globe and Mail*: "Canada is the anti-superpower: the gentle giant that doesn't wield its oil clout as a geopolitical club (think Russia or Venezuela), or set a benchmark for world prices (like Saudi Arabia). It isn't lawless or war ravaged (Nigeria or Iraq)."[147] But a kinder, gentler superpower, Hester would no doubt argue, is no superpower at all. In any case, aspiring to be an energy superpower may not be in line with what most Canadians want for the country. As pollster Michael Adams has concluded, most of us have "a low tolerance for macho-brashness."[148]

Rather than exhibiting the characteristics of an energy superpower, Canada really functions more as an energy satellite. We do not have a made-in-Canada energy strategy designed to serve the needs of our own population first. Indeed, decisions about how Alberta's oil and natural gas resources are distributed are largely made by the industry itself and increasingly by US market demand, that is, in Houston and Washington, not Ottawa or even Edmonton. And, as is discussed in Chapter 4, measures are now being taken through the Security and Prosperity Partnership to ensure that Canada's energy resources are shared with the rest of North America, which is code for the United States.

At the 2007 APEC Summit in Sidney, Australia, Prime Minister Harper promoted Canada as a "clean energy superpower." In many ways, this is an oxymoron. After all, the oil that is produced from the bitumen in the tar sands is still the dirtiest from of crude to be found anywhere in the world. As we shall see in Chapter 5, the production of tar sands crude spews three times the amount of greenhouse gases into the atmosphere as does conventional oil production. In an age of global warming and peak oil, the world does not need an energy superpower. What it needs is a Canada that is a responsible energy producer. This means developing an energy strategy that uses our fossil fuel assets as leverage to make the transition to an economy based on renewable energy sources, which is imperative given the emerging threats of climate chaos and peak oil.

But the Harper government has no intention of moving in this direction. Fortified by its close working relationship with the Alberta oil patch, the Harper team seems to be committed to sticking to its game plan of providing firewall protection for the tar sands, the cornerstone of its vision of Canada as an energy superpower. During the spring of 2008, the two front-runners in the Democratic Party primaries for the US presidency, Senators Hillary Clinton and Barack Obama, called for the reopening of NAFTA. Instead of taking advantage of this opportunity to call for the dismantling of the proportional energy sharing rule in NAFTA, Harper joined with President Bush and President Calderón of Mexico in lauding the benefits of NAFTA when the three leaders met at their annual tri-national summit in New Orleans in April 2008. Indeed, Harper flexed his superpower muscles on that occasion by reminding Americans that "Canada is the biggest and most stable supplier of energy to the United States in the world," and that energy security "is more important now than it was twenty years ago when NAFTA was negotiated, and will be even more important in the future."[149]

It may well be that Harper's depiction of Canada as an energy superpower amounts to no more than political rhetoric, primarily designed to broaden his base of support in the country. Indeed, Canada's case for status as an emerging energy superpower seems to be more delusion than reality — especially when we take a closer look at Canada's role as an energy satellite of the US, and the profound ecological and social catastrophes emerging in the Alberta tar sands.

CHAPTER 4

Fuelling America

For the most part, Canadians are well aware of the fact that the United States is the largest consumer of oil in the world. Many believe that the US gets much of the oil it needs from the Middle East, especially Saudi Arabia. After all, it is often assumed that the need to secure access to Middle Eastern oil was behind the US invasion of Iraq. Ever since the US reached its domestic oil production peak in the early 1970s, securing control over foreign supplies of oil has increasingly become a national priority. Indeed, it is a matter of national security for the US and its survival as the centre of industrial civilization. Being the world's leading military superpower, the US is in a position to marshal its extensive navy, air force and army to secure access to, and control over, the remaining supplies of oil on the planet.

Yet, what most people in this country do not know is that Canada has become the United States' leading fuel pump. In 2004, Canada surpassed Saudi Arabia as the number one foreign exporter of oil and natural gas to the US. In fact, this may be the best kept secret, not only in Canada but in the US as well. Compared to some of the oil-producing countries of the Middle East and elsewhere, Canada, and the Alberta tar sands in particular, undoubtedly seems like a safer and more secure source of petroleum for the US. But, Canada's new-found role as the US's fuel pump also means that Canada has become an energy satellite of the US. What's more, this will have profound implications for our destiny as a nation in the twenty-first century, especially in an age of peak oil, climate change and natural resource depletion.

America's Addiction
In January 2006, George W. Bush declared in his State of the Union address that "America is addicted to oil." Speaking to the annual assembly of the Senate, the House of Representatives, the Cabinet and the US Supreme Court, this was no off-the-cuff remark by the President. It was part of a carefully crafted speech that was televised on all the major US networks. It also struck a nerve in the American psyche, becoming the headline in many of the news reports that followed the next day. Picking up on the metaphor and pursuing the path of Alcoholics Anonymous, some commentators went on to suggest that AA's twelve-step program might help the US rid itself of its addiction to oil.

The US's addiction to oil is rooted in the fact that the United States is the world's leading example of an oil-based industrialized society. Although oil revenues from extraction and refining only account for roughly 5 percent of the US's gross domestic product, oil is essential to a host of other industries that constitute the core of the US economy. Essential to the auto industry, and roads and highway construction, oil is also of pivotal importance to manufacturing, agriculture, aviation and the petrochemical, plastics and rubber industries. It fuels tourism, the home heating business, and suburban commerce. It is also imperative for military operations. In short, the United States' prosperity and way of life is dependent on having access to vast quantities of relatively inexpensive petroleum. The same, of course, applies to the Canadian way of life, which has been largely patterned after that of the US for most of the past century.

As with any other addiction, the addict has a compulsion to continually take the substance in ever-increasing amounts in order to achieve the desired effects. Throughout the twentieth century, the automobile culture, with its vast network of highways, dominated much of the US's economy and demanded increasing volumes of petroleum. But beyond the auto industry, the United States became the epicentre of an oil-based industrial civilization. Spurred on, for example, by new technologies for the production of synthetics and plastics, there was a sudden leap in the demand for petroleum. Similarly, new technologies for pesticides and fertilizers used in the production of new and expanding crops in industrialized agriculture required vast quantities of petroleum. The same was true for the building of a largely military-based economy during the Cold War era of the last century. As a result, the US became increasingly dependent on the availability of almost endless supplies of relatively cheap oil to fuel its economy and society.

Initially, the United States was well endowed with oil. Amongst the major industrializing countries of the nineteenth and twentieth centuries, the US appeared to have the greatest reserves of fossil fuels, especially petroleum. It was simply awash with oil. According to the US Geological Survey, the ground beneath the US originally contained an estimated 345 billion barrels of oil. Between 1859 and 1995, approximately 171 billion barrels were consumed, leaving 174 billion barrels.[150] Of this amount, however, only 32 billion barrels are in known reserves, while sources for the remaining 142 billion barrels are theoretical.[151] For almost three-quarters of the twentieth century, the US saw itself as the world's number one producer, consumer and exporter of oil. Until the 1940s, the US possessed abundant supplies of untapped oil sources to meet its rapidly growing industrial needs. It was not until after World War II that geologists began to send out warning signals about the limits to the US's domestic petroleum supplies.

During this period, the lion's share of world oil production was concentrated in the hands of the so-called "seven sisters," petroleum giants Exxon, British Petroleum, Shell Oil, Texaco, Chevron, Gulf and Mobil. In September 1960, however, five major oil-exporting countries (Iran, Iraq, Saudi Arabia, Kuwait and Venezuela) met together in Baghdad and formed the Organization of Petroleum Exporting Countries (OPEC). Five years later, eight more oil-producing countries joined the OPEC cartel. What distinguished them from the "seven sisters" was that these countries had all nationalized their oil. Through this process, Saudi Arabia and Iraq soon emerged as the world's largest producers and exporters of oil. Although the US and the "seven sisters" largely ignored OPEC throughout the 1960s, it suddenly caught the attention of Washington and the oil majors in the early 1970s when it jacked up its prices for oil exports beyond the standard US base price of 1.80 to 2.00 US dollars per barrel.

Then the five-month embargo following the Yom Kippur War between Israel and its Arab neighbours in 1973–74 created a major oil shockwave. Saudi Arabia and its allies suddenly cut off their oil exports to the US. OPEC became a four-letter word as car and truck drivers brawled with each other in long line-ups at US gas stations. Yet, as John Blair shows in his authoritative study, *The Control of Oil*, the "seven sisters" actually collaborated with OPEC in order to engineer a dramatic price increase during this period.[152] Although the OPEC embargo did not last long, it certainly brought an end to the glut that had plagued the industry during the early 1970s. As oil supplies tightened, Washington discovered that the US was no longer able to use its own reserves to take up the slack: US reserves depleted rapidly under the pressures of the embargo. Already battered by a brewing scandal over Watergate, then US president Richard Nixon found himself thrown into the highly vulnerable situation of having to beg foreign leaders for emergency oil supplies to offset the embargo.

As recently declassified British documents show, the Nixon administration was contemplating a military invasion of Saudi Arabia and Kuwait to seize their oil fields and impose "regime change" in response to the embargo, had it lasted.[153] Instead, diplomacy won out and the embargo ended in March 1974. The whole dramatic experience, however, had shaken up officials in Washington to the point where they seriously took stock of US oil and gas supplies. As it turned out, King Hubbert's predictions about the US's oil supplies were right on the mark. US oil had peaked in 1970. Washington would have to start depending more and more on other countries to fuel its oil-based economy, including its military-industrial complex. In the meantime, the Nixon administration introduced its own package of energy conservation measures that included reduced highway speeds, increased fuel-efficiency standards for automobiles, new insulation requirements in building codes and incentives for alternative energy development.

When Jimmy Carter succeeded Nixon and Ford as president in 1976, he brought forward an ambitious energy plan for the US. Clad in a sweater to emphasize the need to reduce energy consumption during the winter, Carter conducted a series of nationally televised talks with the American people in 1979. Declaring that "hyper-dependence on oil is a deadly trap" that could only be averted by major changes in peoples' lifestyles, and that ensuring the US's energy security was the "moral equivalent of war," Carter went on to say:

To give us energy security, I am asking for the
most massive peacetime commitment of funds
and resources in our nation's history to develop
America's own alternative sources of fuel — from
coal, from shale, from plant products for gasohol,
from unconventional gas, from the sun. I'm proposing a bold conservation program to involve
every state, every county and city and every average American in our energy battle.

Carter's plan to wean the US from its addiction to oil was short
lived. In the media and elsewhere, Carter was ridiculed by Big Oil
and its allies and was later swept out of office by what many
called his naiveté. In April 1980, the price of oil reached an all-
time high, due mainly to the uprising against the US-backed
Shah of Iran.[154] Following this 1980 spike, oil prices went into a
free fall. New oil discoveries were made in the North Sea and later
in Alaska, which quickly allayed the dark memories of the 1970s
oil shocks. As a result, the OPEC producers lost their punch. Desperate for cash, OPEC opened up its pipelines and oil prices took
a nosedive to near record lows. A new era of cheap oil dawned
and gas-guzzling SUVs became the pre-eminent vehicle on the
highways of North America. During the succeeding presidencies
of Ronald Reagan, George H. Bush and Bill Clinton over the next
two decades, the US's oil addiction continued unabated. The
world pumped more oil than was replaced by new discoveries,
and little serious consideration was given to developing either
energy conservation measures or alternative power sources.

By the turn of the century, the United States had come to
depend on oil imports for over 50 percent of its increasing petroleum needs. In 1990, imports accounted for 42 percent of the
US's total annual oil supply. By 1997, the import share had
reached 49 percent. Then, in April 1998, US dependence on

imported oil crossed the 50 percent threshold. This was a long way from the 10 percent dependency on foreign oil imports in the 1950s. Within a few short decades, the US had gone from a country that was relatively self-sufficient to one that had become highly dependent on foreign oil sources to maintain its vibrant economy and enhance its industrial way of life.

When George W. Bush was sworn in as the US President in January 2001, almost ten months before the 9/11 attacks on the World Trade Center and the Pentagon, it was already clear that one of his major priorities would be tackling the US's "energy crisis." From the standpoint of the petroleum industry, the new Bush White House was well positioned to take on this agenda. Not only had George Bush himself been directly involved with the Houston oil patch, but his father has had intimate ties with the Royal House of Saud through Prince Bandar, Saudi Arabia's long-time ambassador to the US, and even the family of Osama Bin Laden himself, the alleged mastermind behind the 9/11 attacks. The Bush family is also closely tied to the Carlyle Group, which specializes in pooling capital for strategic investments in the petroleum industry around the world. Moreover, Bush's cabinet was composed of several people with direct links to the US petroleum industry, including Vice-president Dick Cheney, former CEO of Halliburton Oil Services, and National Security Advisor Condoleezza Rice, a former director of Chevron Oil, who later became US Secretary of State.

Energy quickly emerged as the cornerstone of the new Bush administration, linking foreign policy, defence and domestic policy interests.[155] Nine days after being sworn into office, Bush established the National Energy Policy Development Group (NEPDG) and promptly appointed Vice-president Cheney to head it. From his days as Halliburton's CEO, Cheney had close ties with senior officials in several of the country's major petroleum companies, notably Enron (which was later publicly disgraced for corruption and fiscal malpractice). Among Cheney's colleagues in cabinet were others with close connections to Enron, including Secretary of the Army Thomas E. White, who had been vice-chair of Enron Energy Services, and US Trade Representative Robert Zoellick, who had been a consultant on Enron's payroll. As part of his work on the Cheney task force, Secretary of Energy Spencer Abraham met with as many as 109 representatives of major energy companies between January and May 2001, including Enron, ExxonMobil and ChevronTexaco, all of whom had been big contributors to Bush's 2000 election campaign.

Cheney's mandate for the NEPDG was to develop a long-range strategy to meet US energy needs in the twenty-first century. From the outset, there were two widely divergent paths to choose between. On the one hand, the NEPDG could propose that the US continue on the path of consuming more and more oil and other fossil fuels and therefore become increasingly dependent on imports, recognizing an ongoing decline in its own domestic reserves. Or, on the other hand, the NEPDG could call for a bold turnaround in US energy policy, by outlining a plan for gradual but continuous reductions in petroleum consumption, through effective conservation measures coupled with the development of renewable energy options, thereby reducing its dependency on foreign sources. Despite the Bush administration's close ties to the petroleum industry, the NEPDG claims to have wrestled with these two competing paths before coming up with its recommendations.

Two weeks before the NEPDG report was to be released, Cheney tipped his hand while speaking with reporters at an Associated Press meeting in Toronto. He dismissed the idea that "we could simply conserve or ration our way out" of an energy crisis.[156] Then he went on to say that "conservation may be a sign of personal virtue, but it is not a sufficient basis for a sound, comprehensive energy policy." Instead, he argued that oil would

122 TAR SANDS SHOWDOWN

continue to be the US's prime energy source for "years down the road," and the main objective was to increase access to petroleum sources around the world. Cheney's remarks triggered a barrage of criticism from environmental networks and from Democrats in Congress. In response, the NEPDG revised its report, inserting a much stronger message about the need for conservation, efficiency and alternative energy developments.

But as it turned out, the NEPDG Report, better known as the Cheney Report, was a masterpiece of deception. Released on May 17, 2001, Bush crowned it as the National Energy Policy and promoted it as a green plan that would put priority on the development of renewable energy, including hydrogen-powered fuel cells and biomass fuel sources, along with new technologies and programs emphasizing conservation and efficiency. A closer look at the Cheney Report, however, shows that there is no plan to reduce the US's oil consumption. On the contrary, it appears to be sanguine about continuing increases in annual US consumption rates. While the report does recognize the problem of increasing dependence of foreign oil imports, it proposes that this reliance be reduced not through reducing consumption but through intensifying the development of the US's remaining oil reserves. With the exception of the Arctic National Wildlife Refuge (ANWR) in Alaska, all untapped domestic oil reserves in wilderness areas of the country were to be opened up for exploitation.

In effect, the Cheney Report became the US's recipe for maintaining and increasing its oil addiction. The final chapter of the report, "Strengthening Global Alliances," makes it clear that the basic strategy behind the Cheney energy plan is to increase, not decrease, petroleum imports. To that end, the Cheney Report calls on the President to "make energy security a priority of our trade and foreign policy." Although the report is vague on how much imported oil would be needed in the years to come,

Michael Klare, author of several books on US oil policy and military operations, has made a set of calculations based on the report's charts portraying oil consumption versus domestic oil production. Between 2000 and 2020, says Klare, domestic oil production was expected to decline 18 percent (from 8.5 to 7 million barrels per day), while total oil consumption would grow by 31 percent (from 19.5 to 25.5 million barrels per day). Calculating the difference between the two, says Klare, it's clear that US oil imports would have to rise by 68 percent (from 11 to 18.5 million barrels a day) between 2000 and 2020 to match consumption trends.

The unquestioned dynamic factor in the Cheney plan for the US's energy future is, of course, ever-increasing oil consumption. Apparently the operating assumption behind the report is that American society has an oil addiction that must be constantly fed, not treated. According to the long-range forecasts of the US Department of Energy, total US oil consumption is predicted to rise from an average of 19.7 million barrels a day in 2001 to 28.3 million barrels a day by 2025, a leap of 44 percent.[157] Moreover, the main driving force behind this upward spiral in consumption is the transportation sector. In the US, transportation accounts for two-thirds of all petroleum use. Between 2001 and 2025, the Energy Department predicts that oil consumption for transportation will rise from 13.5 million barrels a day to 20.7 million barrels a day: US oil consumption for transportation will amount to one-sixth of the world's entire oil supply by 2025.[158]

Little wonder that traders on world oil markets keep a close eye on oil consumption patterns in the US. As petroleum analyst Paul Roberts put it, "the colossal U.S. consumption" affects everything in world oil markets. "The most important day of the week for oil traders anywhere in the world," he says, "is Wednesday, when the US Department of Energy releases its weekly figures on American oil use." It is then that "the market makes up its mind whether to be bearish or bullish.… the world's oil players watch the American oil market as attentively as palace physicians once attended the royal bowels: every hour of every day. [Every sign,] … from a shift in energy policy to a trend towards smaller cars to an unusually mild winter," is factored into the equation.[159]

This "colossal US consumption" is proof that the US's oil addiction continues unabated. Admitting, as President Bush did in his 2006 State of the Union address, that "America is addicted to oil" may be the first of AA's twelve steps. But, to "wean ourselves off of petroleum," as Bush declared in the same nationally televised speech, requires taking quite another step altogether. As Richard Heinberg points out, while commenting on Bush's now infamous admission, the problem of "addiction" cannot be confronted without also dealing with the problem of "dependency."[160] Transferring a country's dependency on domestic oil to a dependency on imported oil while, at the same time, increasing oil consumption across the board, merely serves to perpetuate the addiction.

Meanwhile, the 2008 US presidential race may signal a change of direction in US energy policy. Both Republican contender John McCain and Democrat Barack Obama have outlined their own plans for breaking US dependency on imported oil: McCain through his Lexington Project, and Obama via his New Energy for America program. Although both plans have their limitations, Obama's offers the best hope for confronting the problem of America's oil addiction. However, developing a viable US energy policy to reduce dependence on foreign oil imports is now further entangled in, and complicated by, matters of national security and international trade regimes.

National Security
To fuel its oil addiction, the US increasingly depends on foreign oil imports. Indeed, this dependency has been growing steadily. It

is one thing to depend on foreign imports for 10, 20 or even 30 percent of your oil consumption. It is quite another matter to cross the 50 percent threshold, especially since oil is the life blood of this high-tech industrial age in which the US still reigns supreme, albeit haltingly. Securing sufficient supplies of petroleum from other countries has become a matter of national survival. Indeed, oil is understood to be a matter of utmost importance to national security in the United States.[161]

Today, the US depends on foreign imports for over 58 percent of its oil needs. This growing dependency, as Michael Klare points out, is also a sign of vulnerability. As the experience of the oil shocks of 1973–74 and 1979–80 show, dependence on imports has made the US vulnerable to supply disruptions, which leads to oil shortages, price hikes and the potential for worldwide recession.[162] Increasing dependence also means a massive transfer of economic resources from domestic to foreign suppliers, involving trillions of dollars, particularly since the worldwide quadrupling of the price of oil in recent years. In addition, this kind of dependence often requires the granting of political favours to the governments of major foreign suppliers in the form of military protection, advanced weapon transfers, or support at the UN, the WTO or the World Bank. And, "worst of all," says Klare, "dependence generates more insecurity abroad by entangling the US in oil wars overseas that stir up political and religious resentment against the US."[163]

George W. Bush was certainly not the first president to declare US oil and energy supplies a matter of national security. Indeed, it was Franklin D. Roosevelt who first proclaimed this during the closing years of World War II. Despite the fact that the US was then the world's number one producer of oil, the Roosevelt administration realized that the accelerated demands of wartime industry was putting a downward pressure on US oil reserves, which could lead to increased imports, thereby posing a threat to

US security in the long run.[164] Following World War II, Presidents Harry S. Truman, Dwight D. Eisenhower, and John F. Kennedy continued the Roosevelt policy of securing foreign oil supply chains through military operations, especially in the Persian Gulf. During this period, succeeding administrations understood increasing dependence on foreign oil to be a threat to US national security, even though petroleum imports were then supplying no more than 20 percent of US consumption.

The prime focus of this national security strategy has been Saudi Arabia. Possessing one-quarter of the world's proven reserves, an estimated 262 billion barrels of oil, plus the capacity for extra production that could be used to compensate for losses of supply from other major producers, Saudi Arabia became a strategic priority. To secure future foreign oil imports, the Roosevelt administration began to make Saudi Arabia an American protectorate in order to establish a permanent US military presence in the Persian Gulf. In February 1943, Roosevelt told Congress that "the defense of Saudi Arabia is vital to the defense of the United States," and proceeded to transfer military equipment to the Saudi kingdom under the Lend Lease Act.[165] In exchange for reliable access to Saudi oil, every US administration for the next quarter century continued this policy of using its military power to protect Saudi Arabia and the Saudi royal family. Without secure access to Saudi oil, the US would not have been able to achieve the enormous economic growth it did during the quarter century following World War II. The same goes for Europe, which has also depended on access to Saudi and other Middle Eastern oil for its industrial development.

Under the Carter Doctrine of 1980, this national security agenda assumed even greater importance. By 1976, foreign imports accounted for more than 40 percent of total US oil consumption, and the percentage was growing. At the same time, as a result of the Iranian revolution of 1978–79 and the Soviet invasion of

Afghanistan, additional military measures were required to safe-guard the flow of oil from the Gulf. In January 1980, President Jimmy Carter told Congress that the Persian Gulf was of "vital interest" to US security and that his administration would deploy "any means necessary, including military force" to ensure that oil kept flowing from the Gulf.[166] Since there were few US military forces present in the region at that time, Carter backed up his proclamation by establishing the Rapid Deployment Joint Task Force to carry out combat operations in the Gulf.

The Carter Doctrine became the basis for subsequent interventions in the region by the Reagan, Bush and Clinton administrations. In January 1983, President Reagan turned Carter's Rapid Deployment Joint Task Force into a Central Command, known as Centcom, making it one of the five regional "unified commands" that govern US combat forces around the world. In 1987, Reagan invoked the Carter Doctrine, and Centcom forces, using US warships, protected Kuwaiti oil tankers from attack during the latter stages of the Iran–Iraq war. Three years later, President Bush Sr. used the Carter Doctrine to justify the intervention by Centcom forces in Saudi Arabia to prevent an attack by Iraqi forces, who were then occupying Kuwait.[167] And President Bill Clinton continued this policy of containment in the region through Operation Southern Watch, with some 5,000 US pilots and support personnel stationed in Saudi bases between 1991 and 2000 to enforce the no-fly zone over southern Iraq.[168]

By the time George W. Bush succeeded Clinton as president in January 2001, oil imports had crossed well over the 50 percent threshold. Six months previously, the eastern states and parts of the Midwest had experienced oil and natural gas shortages while California endured a series of rolling blackouts. In March 2001, President Bush acknowledged the country was facing an energy crisis that threatened national security. "On our present course," warned the Cheney Report, "America 20 years from now will import nearly two of every three barrels of oil — a condition of increased dependency on foreign powers that do not always have America's interest at heart."[169] But, as Michael Klare points out, instead of blazing a new trail, the Cheney task force vigorously endorsed the same old path of increasing dependency on foreign oil imports, with US military power being deployed to protect supply chains.[170]

What's more, Bush reinforced the Cheney energy plan with his own doctrine on national security in September 2002. Issued one year after the events of 9/11, the Bush Doctrine built on the national security proclamations of previous US administrations and went further to declare the US had a right to (a) make use of pre-emptive strikes against potential aggressors; (b) act unilaterally, if necessary, to protect its interests; and (c) ensure its trade policies and practices serve US national security interests. In effect, the Bush Doctrine provided the justification for the US invasion of Iraq in 2003. Although Bush's own national security doctrine was ostensibly crafted to fight the war on terror after 9/11, it basically serves US interests of enhancing energy security by protecting the US's oil supply chains in the Persian Gulf and elsewhere around the world. Under the Bush Doctrine, energy policy, as a matter of national security, also became more deeply interwoven with US trade and foreign policies.

In today's global economy, the US is certainly not the only country struggling for energy security through control over supply. The Persian Gulf and Africa have long been the source of oil for both Europe and Japan as well. More recently, China and India, both of whom are currently going through a process of rapid industrialization, are relying more and more on oil imports. According to the US Department of Energy, over the next twenty-five years China's and India's oil consumption will more than double, from 5.0 to 12.8 million barrels a day in the case of China, and from 2.1 to 5.3 million barrels a day for

India.[171] Indeed, the oil consumption rates of the developing countries of Asia alone (including China and India) are expected to jump from 15 to 32 million barrels a day during this period. Yet, even the combined oil consumption projected for China and India twenty-five years from now is roughly three-quarters of the US's present consumption rate (though the combined population of China and India is at least eight times that of the US). Nevertheless, these developments will greatly intensify the competition between the US and other emerging economic powers for control over secure access to the remaining sources of conventional oil in the world.

Increasingly, the US finds its control over global oil supply lines being challenged by China and, to some extent, India and Russia. To serve its rapidly growing industrial production needs, China now has to import 45 percent of its oil from abroad, while India imports 70 percent. In recent years, China has signed oil contracts with countries in five key regions: the Middle East, Eurasia, Asia-Pacific, Africa and Latin America. Moreover, China and India have military and political accords with Iran, and China has allied with Venezuela, both of which are in conflict with the US. In short, the growth of China's energy competitiveness, coupled with its military alliances, is bound to heighten US national security concerns and intensify conflicts in the major oil-producing regions of the world.

Meanwhile, the US has been rebuilding its military-based economy to fight the war on terror and to secure control over global oil supply chains. Following the end of the Cold War, symbolized by the tearing down of the Berlin Wall in 1989, the US could no longer justify its military role as "the policeman for the free world" against the Soviet Union. During the 1990s, military spending by the US government levelled off as subsidies and lucrative contracts for the defence industry and arms manufacturers were gradually reduced. In 1997, however, the Project for a New American Century began developing plans for rebuilding the US military. At the centre of the Project were figures who would later become key players in George W. Bush's administration, namely, Vice-president Dick Cheney, Defense Secretary Donald Rumsfeld, Deputy Defense Secretary Paul Wolfowitz and National Security Advisor Richard Pearle.[172] In effect, the Bush team was intent on rebuilding the "military-industrial complex" in the US economy that President Dwight Eisenhower warned about during his final year in the Oval Office, some fifty years ago.

From the outset, the Project was designed to "maintain American security and advance American interests in the new century." In September 1999, Bush himself outlined a major defence policy in his campaign for the presidency, in which he identified three ambitious goals: to "renew the bond of trust between the American President and the American military"; to "defend the American people against missiles and terror"; and to "begin creating the military for the next century."[173] The Bush plan was clearly based on the philosophy of the Project for a New American Century and set the stage for the rebuilding and revitalizing of a military-based economy in the US. The attacks of 9/11 and the subsequent declaration of the "War on Terror" breathed new life into the Project. Less than three-quarters of the way through the Bush presidency, military spending by the US government had skyrocketed by 75 percent, from 315 billion in 2001 to well over 500 billion dollars in 2006. When the costs of the Iraq war plus expenditures in new weapons technology are factored in, by the end of 2006 the total annual military expenditures were approaching the one-trillion-dollar mark.

As political scientist Robert Keohane points out, the strategy behind the rebuilding of the US military economy is to gain control over global oil.[174] With the invasion of Iraq in 2003, the US has effectively become the leading "globo-petro-cop." Currently,

Washington has 725 military bases in 132 countries. Through this kind of military reach, the US is able to secure control over global oil supplies by determining the routing of oil pipelines and exercising undisputed control over the sea lanes through which the world's oil is shipped. By establishing its military bases in strategic locations, the US is also able to prop up unsavoury regimes with armaments and credit in order to gain access to oil supply chains, while marginalizing those countries that stand in the way. At the same time, the US further solidifies its control by making strategic investments in oil-rich countries and regions on a geopolitical basis.

At present, the US carries out its role as the "globo-petro-cop" in the five major oil-producing regions of the world that have become geopolitical theatres of confrontation, most notably with China and, to a lesser extent, with Russia.[175] Briefly, the five strategic oil regions are as follows:

The Persian Gulf. Although Saudi Arabia remains the leading oil supplier in the Persian Gulf, the bonus prize is Iraq, which has abundant sources of high-quality oil that is cheap to produce. Indeed, only seventeen of the eighty potential oil fields in Iraq have been tapped to date.[176] When the US and the UK seized control of Iraq's oil fields through the 2003 invasion, oil contracts between China and Iraq were suddenly cancelled, thereby heightening tensions among these powers in the region. Iraq's still-to-be ratified oil law calls for the granting of fifteen- to twenty-year concessions to the big US and UK petroleum corporations, and up to 75 percent repatriation of profits to the parent companies, along with nearly a doubling of production. Even without the petroleum law being ratified, oil giants such as Exxon-Mobil, Shell, Total SA and BP signed sweet, no-bid contracts with Iraq in 2008. Meanwhile, both China and India have increased their military cooperation with Iran in the form of joint exercises and arms deals, as well as long-term oil-supply contracts.[177]

The Caspian Sea. Here, the rich oil and gas reserves of Central Asia have become a major hotspot as the US, Russia and China scramble to gain control of the region's resources and pipeline routes. Backing different pipeline routes, each of these three big powers is vying for support from regional governments and pitting one against the other. The US, for example, is promoting the Baku-Tibilisi-Ceyhan pipeline, designed to bypass both Russia and Iran in order to transport oil out of Azerbaijan to the Turkish Mediterranean port of Ceyhan, which, in turn, is protected by a large US military base located in Incirlik.[178] In this way, the US intends to encircle both Russia and China with regional pro-American regimes (including Washington's recent support for Georgia, in the face of Russian opposition). And the Canadian military, for its part, is protecting one of these pipeline routes through Afghanistan.[179]

The South China Sea. Here we find the US and China competing for control over vital oil supply routes in the Asia–Pacific region. On the one hand, the US has established an alliance with Japan, South Korea, Australia and possibly India. In 2006, the US Navy carried out its most extensive military operations in the region since the Vietnam War, establishing control over ports and sea lanes. On the other hand, China has countered with what has been called its "string of pearls" strategy, which involves a series of alliances with countries that have harbours along the sea route of its oil shipments, including Bangladesh, Burma, Thailand and Cambodia. The Gwadar port in Pakistan, for example, provides China with a strategic transit terminal for its oil imports and a point from which it can monitor US naval operations in the region.[180]

West Central Africa. Countries such as Nigeria, Sudan and Chad have also become battlegrounds for the control of oil by the US military and other global powers.[181] In order to protect its oil interests in Nigeria and Chad, the US has established a military base in Djibouti. The Djibouti base itself is known to be part of the Trans-Sahara Counterterrorism Initiative, involving troops from a number of neighbouring African countries, including Algeria, Chad, Mali, Mauritania and Morocco.[182] In the same region, Sudan has become China's largest oil supplier, while China, in turn, is now Sudan's number one arms supplier. As a result, China has used its seat on the United Nation's Security Council to prevent the UN from enacting sanctions against Sudan for the atrocities in Darfur.

The upper half of Latin America. Although the US has long considered Latin America to be in its political orbit, China has been making significant inroads. Today, the US maintains its strategic interests in all of the continent's oil-producing countries, notably, Venezuela, Colombia, Bolivia and Ecuador. In order to protect its oil and related interests, Washington has established five military bases in key locations in South America (for example, Colombia, Paraguay and Ecuador), in order to monitor activity on that continent.[183] While Venezuela continues to supply oil to the US, tensions between Caracas and Washington continue to fester. And Venezuela has courted Beijing by inviting Chinese oil companies to explore its oil fields and by enabling the construction of an oil pipeline through Colombia to Tumaco on the Pacific coast, allowing shipments of oil bound for China to bypass the Panama Canal.[184]

Nevertheless, the war on terrorism has spawned a vast new, revitalized military-industrial complex that now spans the globe, securing control over foreign oil supplies and policing a massive network of pipelines, refineries, storage depots and shipping lanes. Under the Bush Doctrine of 2002, Washington claims to have the political authority to engage in pre-emptive strikes if it feels they are necessary to protect access to the foreign oil supplies that now fuel the US on a daily basis. As a result, securing control over oil supplies has frequently been associated with war and bloodshed. This is especially true in three of the five oil-producing regions discussed above. What's more, the costs of maintaining this structural dependency on imported oil from around the world through a militarized energy empire are staggering to the US, both economically and politically, particularly given the skyrocketing price of oil. Indeed, the question arises as to just how sustainable the US's growing dependence on militarily protected, imported supplies is.[185]

Energy Satellite

In March 1999, the cover of the *New York Times Magazine* displayed a giant clenched fist painted over the "stars and stripes" of the US flag. The magazine's cover story was an essay by the *Times'* journalist and author Thomas Friedman called "Manifesto for a Fast World." As the epicentre of global capitalism, Friedman argues, the US must use its considerable economic and political power to "embrace its role as enforcer of the capitalist order." In short, declares Friedman, "America can't be afraid to act like the Almighty Superpower that it is." All the more so when the US needs to secure control over the foreign oil that is essential for the survival of the American way of life. "The hidden hand of the market," argues Friedman, "will never work without the hidden fist." And, "the hidden fist that keeps the world safe for Silicon Valley's technologies is called the United States Army, Navy and Marine Corps."

Unlike people in many other parts of the world, Canadians sometimes have difficulty seeing the US as an empire in the modern world. Perhaps the closer one is to the centre of an economic and military superpower, the more difficult it is to see it for what it is. Also, as some observers have noted, the American empire exhibits characteristics that are different from its imperial predecessors. In any case, four years after Friedman's "Manifesto," prior to the US invasion of Iraq in 2003, the *New York Times Magazine* featured another cover story on "The American Empire" by Canadian academic and author Michael Ignatieff.[186] The term "empire," he wrote, "describes that awesome thing that America's becoming, namely, an imperial power ... And, imperial power means enforcing such order as there is in the world and doing so in the American interest." The cover read: "The American Empire (Get Used To It)."

In this context, it should come as no surprise that Canada is an energy satellite of the US. Historically, Canada has served as a resource satellite for both the British and French empires in the eighteenth and nineteenth centuries, supplying furs, fish, coal, timber and minerals. During the latter half of the twentieth century, Canada increasingly became a major resource satellite of the US, supplying timber, hydroelectric power, oil, natural gas and a range of minerals. Today, Canada is the largest foreign supplier of electricity, oil and natural gas to the US. Approximately 90 percent of its imported electric power comes from Canada. And Canada exports more than 68 percent of its oil production to the US, a major leap from 33 percent in 1985. Meanwhile, export of Canada's natural gas to the US has jumped from 25 percent to over 58 percent of production during the same period. In the eyes of Washington, as we shall see, the Alberta tar sands merely consolidate Canada's role as the US's prime energy satellite.

Reluctantly, the Bush administration in Washington has gradually come to the conclusion that it cannot continue to depend exclusively on unfriendly and outright hostile oil-producing nations around the world. To be sure, this is a recipe for energy instability, not energy security. The Saudi regime, for example, is viewed as being increasingly unstable. Moreover, Matthew Simmons, who advises the Bush administration on energy matters, has raised concerns about the validity of the official estimates of the oil reserves in Saudi Arabia.[187] Iraq, once the second largest oil exporter in the Middle East, now finds its pipelines, pumping stations and shipping facilities constantly being sabotaged as the civil war rages on. Furthermore, the majority of the American people now want the US to get out of Iraq. Elsewhere, internal conflicts and wars are brewing in oil hotspots such as the Caspian Sea, Nigeria and the Sudan, while Venezuela under Hugo Chavez continues to threaten to cut off oil supplies to the US altogether. Policing the world through the US military in order to secure oil supply lines is becoming very costly. Cost estimates range between the Pentagon's figure of a billion dollars a year to the Cato Institute's of 70 billion dollars.[188]

By contrast, Canada, and particularly Alberta, appears to be politically serene, stable and secure. For more than seventy years, Alberta has been governed by regimes that are firmly rooted in conservative values and traditions somewhat similar to those of the Bush administration. Until the tar sands boom, however, Alberta was not counted among the major oil-producing regions of the world. As recently as 2002, the prestigious *Oil and Gas Journal* listed Canada near the bottom of the world's oil-producing countries, with 4.8 billion barrels of conventional reserves. By 2003, the picture had dramatically changed. With the the tar sands factored into the equation, the *Oil and Gas Journal* reported their Alberta estimates had jumped to 180 billion barrels. Suddenly, Canada was propelled into the number two spot on the list of oil-producing countries, behind Saudi Arabia and its estimated 262 billion barrels.

After the *Oil and Gas Journal*'s estimates were published, Washington gave the Alberta tar sands a "thumbs up." "Estimates of Canada's recoverable heavy oil reserves are substantial," declared Vice-president Cheney, "their continued development can be a pillar of sustainable North American energy and economic security."[189] The Center for Strategic and International Studies in Washington went on to say: "There's no more secure supplier to the United States than Canada." Moreover, in 2004 the

Canadian Association of Petroleum Producers estimated that there could well be as much as 315 billion barrels of recoverable crude oil in the tar sands, under current economic conditions and given existing technologies. For these and related reasons, Canada, and more specifically, Alberta, was elevated to a position of strategic priority in Washington in the geopolitics of energy security. As Edmonton energy analyst Mar Anielski puts it: "Oil is no longer simply a commodity for the Americans, but a national security issue."[190]

The "discovery" of the tar sands by Washington marked a dramatic shift in US policy on energy security. The tar sands could provide the US with an abundant supply of heavy crude oil for many decades to come. Americans would no longer have to contemplate going down the path of enduring sharp reductions in oil consumption or rationing of scarce supplies. With a friendly, oil-producing country next door, the US could reduce its dependence on the Middle East and other hostile regions to fuel the US. In 2004, Canada surpassed Saudi Arabia as the number one foreign supplier of oil to the US. To be sure, the era of cheap oil was over, since the production of tar sands crude is much more expensive than conventional production from the oil fields of Saudi Arabia and Iraq. But with the price of oil crossing the 40- and 50-dollar-a-barrel mark, tar sands crude became economically viable. By mid-2008, the cost of producing a barrel of synthetic crude from existing tar sands plants was pegged at less than 25 dollars a barrel, while the estimated cost of producing synthetic crude from a new tar sands plant ranged from 60 to 65 dollars a barrel.[191] And the production and export of tar sands crude would not require a massive military buildup halfway around the world to secure supply lines.

Indeed, the legal framework for a continental energy pact was already in place. Five years after the Canada–US Free Trade Agreement (FTA) came into effect in January 1989, the entire proportional sharing rule, which, in effect, makes Canadian exports of oil and other energy resources to the US compulsory, was incorporated into the North American Free Trade Agreement (NAFTA). As energy specialist John Dillon points out, the proportional sharing rule appears twice in NAFTA, once as Article 315 and again as Article 605.[192] Essentially, Article 315 ensures that if Canada were to take measures that reduced the availability of crude oil or natural gas, for example, Canada would be obligated to make the same proportion of the total supplies of these energy goods available to the US as it had in the past. Canada must continue the flow of crude oil and natural gas to the US in the same proportion of its total supply as the average of the previous three years. Article 605 reinforces this provision by requiring that energy exporting companies, such as are operating in the tar sands, maintain "normal proportions among specific energy ... goods," that is, not substitute, for example, a heavy grade of crude oil for a lighter grade. Taken together, these two NAFTA articles effectively make Canada's energy exports to the US compulsory. However, the proportional sharing rule (really a misnomer) does not apply to the US's other NAFTA partner, Mexico, which negotiated wisely and successfully for an exemption from these constraints on its energy sovereignty.

In a 2008 report, Gordon Laxer and John Dillon show how the NAFTA proportionality rule puts handcuffs on Canada's ability to develop effective strategies to meet our own energy and environmental obligations as a nation.[193] Here, Laxer and Dillon outline three scenarios based on the average level of exports during the last three years for which data are available (i.e., 2004 to 2006). In the first scenario, if Canada were to invoke national energy conservation measures that involved a 10 percent reduction in domestic oil production, we would still be obligated under NAFTA's proportionality rule to export at least 47.5 percent of our total supply of oil to the US.[194] In the second scenario, if Canada were to reserve a portion of its natural gas for the petrochemical industry, or if Alberta were to invoke its fifteen-year reserve rule, we would still be obligated to maintain exports to the US of 51.5 percent of our total natural gas supplies, thereby generating a domestic shortfall in natural gas.[195] In the third scenario, should a major worldwide oil supply disruption require Ottawa to ensure that Quebec and the Atlantic provinces had sufficient supplies by substituting oil from western Canadian and from Newfoundland for imports from the Middle East, Canada would experience a shortfall in its annual domestic supply (31 million barrels) because of its obligation to maintain export levels to the US.[196]

What's more, NAFTA provides US oil corporations with their own enforcement mechanisms. Until NAFTA came into effect, trade agreements only involved nation states and the enforcement mechanism was state-to-state trade sanctions. But, NAFTA also includes an investor–state mechanism, which is outlined in its Chapter 11 on investment. Essentially, Chapter 11 provides a mechanism for foreign-based corporations to sue the governments of member countries in which they are operating for alleged violations of the NAFTA rules. Cases are adjudicated by NAFTA tribunals, appointed by the three member governments, in which the rules of NAFTA take precedence over the domestic laws of the country involved in the suit. So, for example, if Alberta were to face serious shortages in natural gas supplies due in large measure to the increasing amounts consumed by the tar sands industry, it could invoke the province's own law that reserves must not be allowed to fall below the amount needed to supply Albertans for fifteen years. If Alberta tried to enforce this rule, either by rationing use in tar sands production or by curbing US exports, then the US oil corporations affected by these protective measures would be in a position to sue Ottawa for violation of the NAFTA proportionality rule.[197] In this way, NAFTA provides the US petroleum industry with a powerful set of tools to determine, if not dictate, Canada's energy policy decisions. ExxonMobil's 2007 decision to challenge the Newfoundland government's regulation of its Hibernia and Terra Oils operations under Chapter 11 of NAFTA may open the door to more chilling legal actions along these lines.[198]

As if to further establish Canada as the supply satellite in its continental energy security strategy, President Bush signed the United States Energy Policy Act into law in August 2005, calling it America's "energy strategy for the twenty-first century." The new law requires that a continental plan be developed to make North America energy self-sufficient by the year 2025. The cornerstone of that plan is the Alberta tar sands. Although the Energy Policy Act calls for a reduction of one million barrels a day in US oil consumption, this reduction does not begin until 2015, a time when there's expected to be a surge in tar sands production and exports. In order to develop this continental energy plan, the Act established a United States Commission on North American Energy Freedom, composed of "qualified citizens from Canada, Mexico and the United States." This Commission is to submit a report to Congress every year with recommendations on what is to be done to ensure that the goals of "North American energy freedom" are achieved. In effect, the US Congress would determine, based on the US's interests, an energy security plan for all of North America.[199]

The US Energy Policy Act of 2005 contains at least two provisions that reaffirm Canada's status as an energy satellite of the US.[200] First, in a section called "Use of Fuel to Meet Department of Defense Needs," the law specifically designates tar sands production to serve the fuel needs of the US military. At the discretion of the US Secretary of Defense, the law states that as much crude oil as possible from the tar sands will be processed in refineries south of the border for purposes of fuelling the

American military. Second, in a section labelled "Partnerships," the Energy Policy Act calls for a special relationship or partnership to be developed with the Province of Alberta, Canada, "for purposes of sharing information relating to the development and production of oil from tar sands." Note, this would be a direct, bilateral partnership between Washington and the Alberta government, excluding the Government of Canada.

Although Bush's Energy Policy Act was followed six months later by his infamous "America is addicted to oil" statement in the 2006 State of the Union address, US energy secretary Samuel Bodman promptly reassured the American public that the US government would not be imposing harsh energy conservation measures to reduce oil consumption. In fact, oil consumption in the US was rising, not declining. In 2005, when the new energy law was enacted, the US recorded its largest increase in oil consumption in more than twenty years.[201] Instead, the Bush administration's strategy, said Bodman, was to substantially reduce its dependence on foreign oil imports from politically unstable regions of the world. More specifically, the objective was to reduce imports from unstable oil-producing regions such as the Persian Gulf by 75 percent by 2025. In Bodman's view, the oil import gap would be made up by Canada via rapidly increasing crude production from the tar sands.[202] In December 2007, however, Congress passed another law, the US Energy Independence and Security Act, which may have put the brakes on the Bush administration's plans by forbidding the use of dirty fuels such as tar sands crude by government agencies, including the Pentagon.[203] (See Chapter 5 for further discussion of this act.)

In the meantime, Washington was making moves to ensure that China would not gain access to the Alberta tar sands. In 2005, when the energy policy bill was being debated in Congress, a move by the state-owned China National Offshore Oil Corporation to counter a bid by Chevron for the purchase of Unocal, a California-based company, was blocked overwhelmingly by a vote of 398 to 15 in the House of Representatives. Viewed as a threat "to impair the national security of the United States," China's national oil company was forced to withdraw its bid around the same time as the energy bill became law. US officials continue to closely monitor ongoing talks between Chinese and Canadian oil companies. China, for example, through PetroChina International Co., a subsidiary of state-owned China National Oil and Gas Exploration and Development Corp., has signed a memorandum of understanding with Enbridge Pipelines to construct a pipeline from Edmonton to Kitimat to transport oil from the tar sands to the Pacific Coast.[204] And Syncrude has already experimented with trial shipments of oil to PetroChina, which has raised eyebrows in Washington.

In late 2005, the *New York Times* ran a feature article entitled "China Emerging as Rival to US for Oil in Canada." Although, from the standpoint of market leverage, it is to Canada's advantage to have both the Chinese and the Americans actively competing for access to tar sands oil, China's participation is seen as a provocation by Washington. "China's thirst for oil," reads the opening paragraph of the *Times* piece, "has brought it to the doorstep of the United States."[205] Since US-based companies control between 40 and 50 percent of the investment in Alberta's oil production, the basic message of the article is that it's the US, not China, that has proprietary interest. In other words, Alberta is the US's energy satellite. And, if China strikes more deals with Canadian companies possessing significant leases in the tar sands, and if the Enbridge pipeline to Kitimat is built, pressures from Washington are bound to increase.

Following the passage of its Energy Policy Act, US officials took matters in hand. Immediately after the election of Stephen Harper as Prime Minister, in January 2006, a two-day summit organized by the US Department of Energy and Natural

Resources Canada was held in Houston, Texas. There, leading industry and government officials from both countries pledged to rapidly increase tar sands production from one million to five million barrels of oil per day by around 2020. If the tar sands were producing five million bpd today, Canada would be supplying one-quarter of all oil consumed in the US and close to half of all American imports of oil would be from Canada, assuming all of the tar sands crude produced was exported. To meet this export target, the US expects Canada and Alberta to relax their already limited environmental and social regulations on energy mega-developments. As CBC News reported, the minutes of the Houston Summit make it clear that Canada would have to "streamline" its environmental regulations for new energy projects, if it is to increase its output by 500 percent so quickly. A year later, the Harper government established the Major Projects Management Office in its 2007 budget with the intention of "cutting the average regulatory review period [for large natural resource projects] from four years to two years." Moreover, the American Petroleum Institute made it clear that the five-fold increase in oil exports to the US from Canada would require the immediate construction of additional pipeline capacity between the two countries.

Shortly after the Houston meeting, US Energy Secretary Bodman launched a speaking tour to woo Canadian political and economic elites. It began in March 2006 with a speech at the Canadian embassy in Washington where he emphasized how important Canada is to the new US energy plan, declaring: "We are certainly very anxious that the oil sands development be as swift as possible."206 In July 2006, Bodman took his entourage north to Fort McMurray, where he flew over the strip mines in his platoon of five helicopters, enjoyed a photo op with his host Alberta Premier Ralph Klein and took a ride on one of the giant 400-ton yellow trucks. Later, at a luncheon meeting with oil executives, Bodman reiterated Washington's commitment to reduce its dependence on oil imports from "unstable regions by five million barrels a day by 2025," and then stated categorically: "No single thing can do more to help us reach that goal than realizing the potential of the oil sands in Alberta."[207] (Figure 4.1)

In many ways, Bodman was showcasing Alberta, and the tar sands in particular, as the US's energy satellite. What made Alberta worth showing off to the rest of the world, of course, was the "discovery" of the massive new reserves in the tar sands that had suddenly propelled Canada into second place amongst the world's oil-producing regions. Yet, the building blocks for this newly minted role as energy satellite had already been cemented and sealed a decade and a half earlier through the Canada–US Free Trade Agreement and subsequently NAFTA. The proportional sharing clause would virtually guarantee the US would have a continuous and increasing flow of crude oil from Canada to meet a substantial portion of its oil consumption demands. As the US petroleum giants invest more and more into the development of the tar sands, any future moves by governments north of the border to put a quota on oil exports or constrain the industry's operations could be checkmated under Chapter 11, which permits these corporations to sue the Canadian government for violation of any of the rules in NAFTA.

As if these measures were not sufficient, Canada's status as the US's energy satellite is now being further entrenched through the Security and Prosperity Partnership. Ever since NAFTA's implementation in January 1994, a small group of academics, former government officials and business leaders have pushed for deeper continental integration by advocating a "NAFTA Plus" agenda. In 2003, the Canadian Council of Chief Executives (CCCE) — formerly known as the Business Council on National Issues (BCNI) — launched its own "North American Security and Prosperity Initiative." Led by Rick George, president of Suncor, a

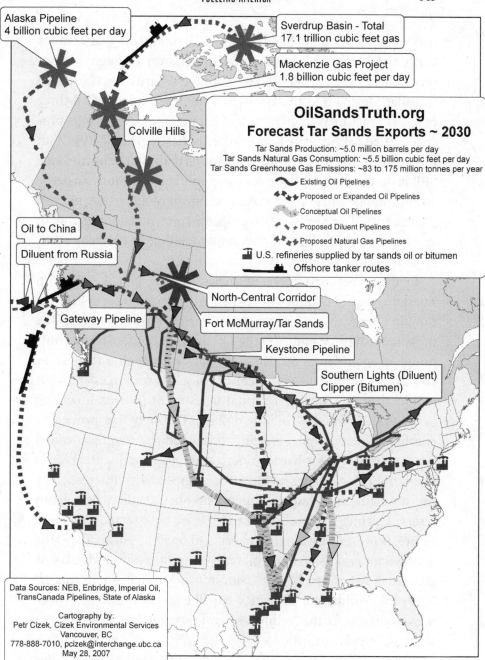

Alaska Pipeline
4 billion cubic feet per day

Sverdrup Basin - Total
17.1 trillion cubic feet gas

Mackenzie Gas Project
1.8 billion cubic feet per day

Colville Hills

OilSandsTruth.org
Forecast Tar Sands Exports ~ 2030
Tar Sands Production: ~5.0 million barrels per day
Tar Sands Natural Gas Consumption: ~5.5 billion cubic feet per day
Tar Sands Greenhouse Gas Emissions: ~83 to 175 million tonnes per year

Existing Oil Pipelines
Proposed or Expanded Oil Pipelines
Conceptual Oil Pipelines
Proposed Diluent Pipelines
Proposed Natural Gas Pipelines
U.S. refineries supplied by tar sands oil or bitumen
Offshore tanker routes

Oil to China

Diluent from Russia

North-Central Corridor

Gateway Pipeline

Fort McMurray/Tar Sands

Keystone Pipeline

Southern Lights (Diluent)
Clipper (Bitumen)

Data Sources: NEB, Enbridge, Imperial Oil,
TransCanada Pipelines, State of Alaska

Cartography by:
Petr Cizek, Cizek Environmental Services
Vancouver, BC
778-888-7010, pcizek@interchange.ubc.ca
May 28, 2007

Figure 4.1. Tar Sands Exports and Pipeline Corridors in 2030. Courtesy Petr Cizek.

major player in the tar sands, the CCCE outlined a platform for
further strengthening continental integration through harmo-
nized energy regulations and a "resource security pact" between
Canada, Mexico and the US.[208] In 2005, the leaders of all three
countries met in Waco, Texas, and launched a NAFTA-plus program called the Security and Prosperity Partnership (SPP). With
minister-led working groups and more than 300 priorities, the
SPP is designed to work below the public radar to further deep-
en continental integration and the creation of a North American
community. The SPP is not transparent nor is it accountable to
any democratically elected government, either in Canada, Mexico or the United States.

To date, the centrepiece of the SPP is its North American
Energy Working Group, whose mandate is to enhance collaboration on energy resources and streamline energy regulations. One
of its key advisory bodies is the Oil Sands Experts Group, composed of petroleum industry officials whose objective is "to
collaborate on the development of oil sands resources" in what
is now the world's largest capital investment project, involving
between 125 and 150 billion dollars. Through this process, a
continental energy plan is being developed to provide so-called
"energy freedom" for North America. Yet, this continental plan is
primarily oriented to serving US energy security interests. For
the moment, it is largely dependent on Canadian petroleum
reserves, principally tar sands crude, as we have seen. In the near
future, however, Washington expects to open up Mexico's petroleum sector and its offshore oil reserves in the Gulf of Mexico. At
the North American Leaders' Summit in Montebello, in August
2007, the mandate of the SPP's North American Energy Working Group was further reinforced and expanded to include these
strategic energy priorities.[209]

In effect, Canada's status as an energy satellite of the US has
been secured through various legal regimes, including NAFTA,

the SPP and US energy legislation.[210] Instead of having a made-in-Canada energy strategy to ensure that our oil, gas and hydroelectric power serves the needs of people and the environment in this country first, our energy resources are being developed primarily to provide exports to the US. Under NAFTA, and by extension the SPP, we no longer have the option of an energy policy that is democratically determined and controlled through elected governments. Instead, it is the US market, the oil industry in Houston and the policy-makers in Washington that now largely determine Canada's energy policies. As a consequence, Canada is moving in the opposite direction of most other major oil-producing nations. While Latin American oil-producing countries ranging from Venezuela and Bolivia to Ecuador, Argentina and Brazil are asserting national sovereignty over their energy resources on behalf of their own people and, in some cases, the environment, Canada has virtually surrendered control over its oil and natural gas resources to US-based petroleum companies. Instead of being the world's next energy superpower, Canada has become little more than an energy colony within the American empire.

CHAPTER 5

Ecological Nightmare

The mad rush to develop the tar sands in order to fuel America's insatiable appetite for oil comes with an enormous environmental price tag. After all, this is not the easy-to-get-at oil that gushes up from the ground or lies in pools just below the surface of the earth. No, this is the hard-to-get-at heavy oil (bitumen) mixed into sand that lies deep down in the sedimentary basin beneath the earth's surface. The only way to get at it is with brute force.

At first, detonating a nuclear bomb was proposed, but when that plan was jettisoned (at least for the time being) giant earth-moving equipment was brought in to do the job. Today, enormous machines mow down trees (and the wildlife they support), roll up hectares upon hectares of muskeg, drain expansive wetlands and reroute entire river systems. Every day, vast quantities of fresh water, vital to the ecosystem of this bioregion, are being depleted or contaminated at an alarming rate in order to get the tar sands out of the ground and to separate the bitumen from it. And, to make matters worse, the tar sands industry spews more carbon emissions into the atmosphere on an increasing scale than virtually any other industry in the Canadian economy.

Global Warming

Most people in Canada today, including many Albertans, are genuinely and increasingly concerned about the realities of climate change. After all, we are a northern people living in close physical and geographical interconnection with one of the most fragile and vulnerable ecosystems of the planet, the Arctic. As the plethora of reports from the United Nations Intergovernmental Panel on Climate Change (UNIPCC) have become public, Canadians have generally responded by demanding governments do more about the greenhouse gas emissions that are heating the planet. Public opinion polls have consistently shown that the Canadian public wants our governments to meet the commitment made by Canada in the Kyoto Protocol to reduce this country's greenhouse gas emissions to 6 percent below 1990 levels. What most Canadians do not know is that the Alberta tar sands are becoming Canada's number one global warming machine.

The forests of northern Alberta, which are being strip-mined for the production of tar sands crude, are part of the boreal ecosystem, known as the northern lungs of the planet. The boreal has the potential to recapture more carbon from greenhouse gas emissions than any other ecosystem in the world, including the tropical rainforests of Africa and South America. The boreal peat lands, soils and trees are a natural storehouse for carbon. From a climate change perspective, therefore, the boreal constitutes one of the most important natural features in the world. Yet, the boreal carbon sink is being ripped away by strip mining and being replaced by a vast network of open pit mines, wells, roads, pipelines and hugely toxic waste ponds. Thousands of hectares of previously untouched boreal are being systematically destroyed.

At one point, the Alberta government was on the cutting edge of reducing greenhouse gas emissions. In 1988, the Energy Efficiency Branch of the Department of Energy produced a report called "A Discussion Paper on the Potential for Reducing CO_2 Emissions in Alberta, 1988–2005." Recognizing that global warming poses a serious threat to life on the planet, the report identified the causes as the accumulation of carbon dioxide, methane, nitrogen oxides and chlorofluorocarbons in the atmosphere. The report went on to outline a plan for reducing Alberta's 2005 greenhouse gas emissions levels to 20 percent below 1988 levels. The plan included a list of some 300 measures for energy conservation. In short, Alberta had a plan of action that, if implemented, would have pursued targets for reducing greenhouse gas emissions well below those set by the Kyoto Protocol.[211]

Alberta's path-breaking report, however, was kept under wraps. In 1990, while former premier Ralph Klein was environment minister, the report was hidden away in the government's archives. After Klein became premier, he terminated the Energy Efficiency Branch and promptly slashed budgets related to renewable energy and environmental protection. Throughout the 1990s, Klein and his ministers openly campaigned against the Kyoto Accord. At one point, Klein jokingly suggested that global warming is caused by dinosaur farts.[212] In collaboration with individuals in the oil industry, members of Klein's Conservative Party and the former Reform and Alliance Parties were instrumental in forming an organization called Friends of Science, which steadfastly denied that global warming is caused by human activities and campaigned vigorously against Kyoto.

Today, it is well known that the energy sector is one of the largest generator of greenhouse gases in Canada. Within the energy sector, the tar sands industry is the single biggest and fastest-growing greenhouse gas emitter. In fact, the emissions from the production of tar sands crude are at least three times greater than those generated by conventional oil production. The reason is that large amounts of another fossil fuel (i.e., natural gas) are used both to extract the bitumen and to upgrade it into synthetic crude oil. According to the Pembina Institute, conventional oil produces, on average, 28.6 kilograms of carbon dioxide per barrel of oil while tar sands oil generates 85.5 kilograms.[213] What this means is that one open pit mine in the tar sands along with one upgrader will emit as much greenhouse gas into the atmosphere in a single day as 1.35 million cars on the road.[214] This is why the tar sands are set to become Canada's number one global warming machine.

Now, as we saw in Chapter 3, tar sands production is expected to grow and multiply several fold over the next decade and beyond. This means that greenhouse gas emissions from the tar sands will also multiply. It is generally agreed that tar sands production in 2005 emitted 37 million tonnes of greenhouse gases (up from 23 million tonnes in 2000). But, with the projected growth in tar sands production between now and 2015, the Pembina Institute predicts that this could increase to 126 million tonnes of greenhouse gases annually, assuming that production continues to be fuelled by natural gas.[215] However, calculations carried out by a team of Swedish energy policy analysts predict an even greater volume of emissions, if the industry moves to using a combination of bitumen residue and natural gas to fuel production, since burning bitumen residue gives off more CO_2 than burning natural gas does.[216] Table 5.1 shows the predictions for each of these scenarios.

Pembina also estimates that the tar sands will account for half the growth in Canada's carbon emissions between 2003 and 2010.[217] Table 5.2, however, provides a more comprehensive picture of the industry's current and projected greenhouse gas emissions. Looking at the figures showing emissions at each stage

Table 5.1		
Greenhouse Gas Emission Scenarios	2005	2015
Greenhouse gas emissions generated by burning of natural gas in tar sands production	37 million tonnes	126 million tonnes
Greenhouse gas emissions generated by burning a mix of bitumen residue and natural gas in tar sands production	—	135–165 million tonnes

Source : Polaris Institute.

of production, it is clear that the tar sands industry could be generating as much as *164 million tonnes* of greenhouse gas emissions by 2015 — closer to the figures that the Swedish team has calculated based on the burning of a mix of bitumen residue and natural gas. Consequently, some now maintain that greenhouse gas emissions from tar sands production are three to five times that of conventional oil.

In any case, by 2010, the tar sands are expected to be the fastest growing contributor to Canada's greenhouse gas emissions of any single industry in the country. Indeed, emissions from the tar sands will rival those of entire countries. According to the World Resources Institute, greenhouse gas emissions from the tar sands alone could soon match the annual output of the Czech Republic, and be twice the output of Peru, three times that of oil-producing Qatar, and ten times that of Costa Rica. And according to a report by the Suzuki Foundation, three of Canada's top five industrial polluters now are from the tar sands industry.[218]

Today, Canada is a long way from meeting its Kyoto commitments. Instead of reducing greenhouse gas emissions to 6 percent below 1990 levels, as required under Kyoto, our total emissions are likely to be 32 percent higher than 1990 levels by 2010. In addition, the Harper government has managed to confuse the public about calculating emission reductions. In its April 2007

Table 5.2

Greenhouse Gas Emissions by Oil Sands Production Processes
Total Emissions (in millions of tonnes CO_2 per year)

2006	2007	2008	2009	2010	2011	2012	2013	2014	2015	2016	2017	2018	2019	2020
Excavating: Mining														
14	16	18	22	28	31	36	41	43	43	44	49	50	52	52
Excavating: In situ, SAGD (oil sands)														
10	13	16	19	23	29	35	39	42	46	47	47	47	47	47
Upgrading: oil sands & extra heavy oils														
24	28	33	39	47	53	61	68	73	76	78	82	83	84	85
TOTAL: excavating and upgrading														
49	57	68	80	97	113	132	148	157	164	168	178	180	182	184

Source: M.C. Herweyer and A. Gupta, 2007.

report, "Turning the Corner," the government outlines targets for emission reductions based on year 2006, rather than the internationally recognized benchmark of 1990. So, although the Harper plan calls for a 20 percent reduction in emissions by 2020, that emission level is still 2 percent higher than 1990 levels — eight years after Kyoto's target date of 2012. At this rate, Canada's emissions level in 2050 will be significantly higher than those required to avoid a global meltdown later this century. Table 5.3 illustrates how off base Canada is currently in reducing carbon emissions in comparison with other G-8 countries.

As long as the tar sands industry is allowed to expand, generating ever-increasing carbon emissions, Canada is bound to remain a major laggard in the fight against global warming.

What's more, the Harperites have adopted intensity-based targets for the reduction of greenhouse gas emissions by industry. For the tar sands, the obligation is to reduce emissions on a per barrel basis. To some extent, the industry is already doing this. However, this regulation does nothing to curb total emissions from an ever-expanding industry such as the tar sands. If Canada applies intensity-based targets to the tar sands, greenhouse gas

Table 5.3		
G-8 Countries' Carbon Footprints		
Country	2006 Greenhouse Gas Emissions (in millions of tonnes)	Change (1990 to 2006)
Russia	2,190.4	– 34.2%
Germany	1,004.8	– 18.2%
Britain	655.8	– 15.0%
France	541.3	– 3.9%
Japan	1,340.1	+ 5.3%
Italy	577.9	+ 9.9%
Canada	720.6	+ 22.0%
United States	7,017.3	+ 32.0%

Source: The Globe and Mail, *July 9, 2008 (A-6).*

emissions will likely multiply three- to four-fold by 2020 given the planned increases in production. In addition, the Harper Conservative government has given the tar sands industry a three-year free ride. Companies operating in the tar sands are exempt from the government's new environmental regulations until 2011.

Alberta's own "climate change strategy" makes matters worse. The new Stelmach government has come up with a set of emissions reduction targets even further below Harper's. Already, Alberta accounts for roughly 35 percent of all of Canada's emissions. While the Harper government has set targets ranging between 50 and 60 percent reductions in greenhouse gas emissions by 2050, the Stelmach government's target for emissions reductions is only 14 percent. According to the National Round Table on the Environment and the Economy, for Canada to reduce its emissions by even 50 percent by 2050, Alberta must reduce its emissions by 40, not 14, percent. There is no way that Canada can meet even the low-ball Harper target if Alberta reduces emissions by only 14 percent. Writes Jeffrey Simpson of the *Globe and Mail,* this will make Alberta "the pariah of jurisdictions internationally."[219]

As the December 2007 International Conference on Climate Change, in Bali, Indonesia, demonstrated, Canada is far behind many other industrialized countries when it comes to reducing greenhouse gas emissions. The entire twenty-four-nation European Union, for example, is committed to 30 percent emission reductions below 1990 levels by 2020, while Canada's targets remain at 2 percent above 1990 levels at that point.

At the same time, the search for a magic bullet continues. One option being touted is called CCS, carbon capture and storage (or sequestration). CCS is a technique used to capture carbon from the production process and store it deep underground. It is the same technology that's being used successfully in Alberta and Saskatchewan to enhance oil and gas recovery by pumping carbon dioxide into conventional wells nearing depletion in order to extend their productivity. By returning and storing carbon dioxide deep underground in old oil and gas fields and salty aquifers, less carbon will be released into the atmosphere. In effect, the theory is to take the smokestacks of the upgraders and petrochemical plants, turn them upside down, and pump the carbon dioxide underground. To construct an integrated CCS network in the tar sands, however, would require a pipeline for connecting and capturing the carbon dioxide emissions from all the various tar sands plants, which may prove to be a complex undertaking.

Carbon capture and storage has serious limitations and raises important questions. While the technology may hold some promise, the National Round Table on the Environment and Economy says that large-scale application of CCS has not been tested and it is fraught with both challenges and risks. Once the carbon dioxide is pumped under ground, there is no guarantee that it will not leak. A large leak of concentrated carbon into the atmosphere could be disastrous to the environment, while a slow leak would defeat the purpose of storing it in the first place. As well, the financial costs are considerable. According to some estimates, building and operating a CCS system in the tar sands designed to capture and store just twenty million tonnes of carbon dioxide a year by 2020, would cost 16 billion dollars over a twenty-year period. To capture and store all of the carbon emissions from the tar sands, says the Intergovernmental Panel on Climate Change (IPCC), would increase energy costs by 40 percent.[220]

In March 2008, federal environment minister John Baird announced that the Harper government would be promoting the development of CCS technology in the tar sands. All companies starting their crude oil operations in the tar sands from 2012 onwards will be required to apply this technology. There are several flaws in the government's plan. Under these new regulations, the companies will have until 2018 to comply, i.e., to ensure that their carbon emissions are reduced through CCS or a combination of other abatement technologies and carbon credits. During the intervening period, however, carbon emissions from the tar sands will be escalating due to the projected three- to four-fold increase in production. Moreover, according to a report from the Energy Resources Group at UC Berkeley, by Alex Farrell and Adam Brandt, only 10 to 20 percent of the tar sands carbon emissions could be captured and sequestered through CCS.[221] Although new technologies may improve CCS, Farrell and Brandt point out that the emissions due to the ultimate combustion of the fuel would be unaffected. On top of this, the Alberta government has promised 2 billion dollars for the development of a CCS system in the tar sands, which falls far short of cost estimates, raising the question of who will pick up the slack — the industry or the provincial government?

Meanwhile, Canada continues to heavily subsidize the tar sands industry in other ways. Two weeks before the Kyoto Protocol was ratified in 2002, Canadian environmentalists challenged the federal government to overhaul its policy of providing lucrative subsidies to the oil sector. As we saw in Chapter 3, the tar sands companies are the recipients of some of the biggest federal subsidies given. After reviewing the resource allowance and the accelerated capital cost allowance for tar sands production, Pembina concluded, in a January 2005 report, that these lucrative federal subsidies are undermining the effort to realize environmental priorities in the region. Two years later, Pembina and the World Wildlife Fund released a joint report on the environmental practices of the major companies involved in the tar sands that gave companies a failing grade on reducing their overall ecological impact significantly and for being slow to invest in new, more environmentally friendly technologies.

The issue of federal subsidies, however, is further complicated by the lingering memories of the 1980 National Energy Program (NEP) by which Ottawa intervened directly in the Alberta oil patch, increasing Canadian ownership and control of the industry and taking a cut of oil revenues. During the debate over Kyoto in the fall of 2002, the Klein government in Alberta joined forces with the oil and gas industry to vigorously campaign against Canada's participation. Ralph Klein called Kyoto "the goofiest, most devastating thing that was ever conceived," and the industry warned that investment would dry up in the oil patch if Canada became a signatory to the treaty. Throughout the debate, comparisons were made in Alberta between Kyoto and the reviled NEP, both of which were alleged to have contributed to the marginalization of the oil industry. But when the Kyoto Protocol was ratified by the Chrétien government, Klein told the *Globe and Mail* that it didn't matter, since enough compromises had been achieved to ensure that "there is no peril to the oil sands."[222]

Even so, the environmental challenge of the tar sands is bound to stir up more regional tensions and divisions. Former Alberta premier Peter Lougheed, who promoted the tar sands and the building of the Syncrude plant, has recently had a change of heart, issuing dire warnings of his own. Anticipating there will be a battle with Ottawa over the future of the tar sands, in the summer of 2007 Lougheed told the Canadian Bar Association that this would be "ten times greater" than the battle over the NEP, which he experienced first-hand in 1980. While later qualifying his remarks in an interview with the *Globe and Mail*, Lougheed emphasized that environmental issues are "sensitive." He went on to say: "The national energy program was looked at as a battle between two jurisdictions over money. If [the next] conflict evolves as it might, on environmental issues, it becomes much more emotional with citizens."[223]

Where such a battle is likely to flare up is over the question of a carbon tax or other serious environmental levy. Rejecting a carbon tax, the Harper government has opted for a Climate Change Technology Fund, calling on companies to pay a fee of 15 dollars per tonne of greenhouse gas emissions to be put into the fund, increasing to 20 dollars between 2010 and 2017. The three opposition parties, however, have called for stiffer penalties, requiring companies to pay fees of 20 to 30 dollars per tonne for emissions above the Kyoto targets set by Canada. And the National Round Table on the Environment and the Economy has gone much further, arguing that fees ranging from 190 and 240 dollars per tonne are required if Canada is to reach even the Harper government's targets of reducing emissions by 50 to 60 percent by 2050.

Stephen Harper wasted no time in going on the attack when Liberal leader Stéphane Dion announced his carbon tax plan in June 2008. In the weeks before the Liberals had even announced the details of their Green Shift plan, the centrepiece of which was to be a carbon tax, the Conservatives were rolling out their own ads mocking it as another Liberal tax-and-spend scheme. Within an hour of Dion's formal announcement of the plan, Harper promptly denounced the proposals. Later, he publicly stated that whereas the NEP was designed to "screw the West," the carbon tax scheme was designed to "screw the rest of the country."

At the same time, there are indications that strong federal actions on the environment could provoke a constitutional debate. Alberta's current premier, Ed Stelmach, has already indicated that he is ready to mount a constitutional challenge if necessary to protect the tar sands. "We'll take every action that's available to us in terms of protecting the constitutional rights of every Albertan." Under the Canadian constitution, the Alberta government has exclusive jurisdiction (as do all the provinces, but not the three resource-rich territories) over its non-renewable resources. At the same time, the federal government has the right to protect the environment of the country through legislation. The Supreme Court of Canada has recognized that Canada does have wide-ranging latitude when it comes to regulating emissions. Even so, legal experts warn that the looming battle between Canada and Alberta over the regulation of greenhouse gases and its impact on the tar sands industry could turn into a major constitutional storm.

Still, the ecological fallout from the tar sands could prove to be the biggest blow to the oil industry's bottom line. In the United States, pressures are building that could significantly affect the sales of tar sands crude — the dirtiest form of oil on the market — in the biggest market that currently exists for this product. In California, there are now laws on the books requiring the use of "low carbon fuel" and imposing levies on the use of "dirty" fuels. California's legislation takes into account the carbon emissions of the whole fuel cycle — extraction, upgrading, transport and refining — which makes tar sands crude very vulnerable. Based on these standards, tar sands crude is viewed as a dirty fuel. In addition, California's Governor Arnold Schwarzenegger has been actively promoting "low carbon fuel" standards across the US, and more than ten other states are enacting similar laws. Schwarzenegger also signed agreements in May 2007 with two provinces, British Columbia and Ontario, to develop and adopt low carbon fuel standards.

Indeed, growing public pressure against dirty fuels in the US could eventually close that market to tar sands crude. In January 2008, President George W. Bush signed a new law, the Energy Independence and Security Act, preventing departments of the federal government from using synthetic crude oil if its production generates more greenhouse gases than the production of conventional oil does. Although there is some dispute about how this will apply to tar sands crude, this law could throw a monkey wrench into the production and export plans of the Alberta government and the tar sands industry. Moreover, the US mayors, meeting in Miami in June 2008, agreed to scrutinize the oil consumption of their cities to ensure that no use is made of dirty fuels such as tar sands crude. At the same time, Democratic presidential nominee Barack Obama pledged to reduce the US appetite for "dirty, dwindling, and dangerously expensive" oil, and his officials have hinted that tar sands crude may not be part of the long-term energy plans of an Obama administration.[224]

As strategically important as these countermeasures are, it remains to be seen whether they are sufficient to slow down, let alone stop, imports from Canada's tar sands to the US. As we have seen, Washington's rapidly growing dependence on Canadian petroleum exports to satisfy the US's insatiable appetite for oil may be far too great to permit a full boycott of dirty tar sands crude. Even so, what is at stake here, in both the short-

and the long-term, is Canada's reputation in the eyes of the world. In an age of climate change, the tar sands are considered to be a dirty business. Like it or not, Canada is no longer seen as an eco-friendly country. As the wave of protests against Canada's environment minister at the 2007 international climate change meetings in Bali and against the prime minister at the 2008 G-8 Summit in Hokkaido, Japan, demonstrated, Canada is getting a reputation as an eco-unfriendly country. This is due, in large measure, to the development of the tar sands. Canada is becoming known as the "tar-nation" of the world.

Water Crisis
The Alberta tar sands are not only Canada's number one global warming machine. They are also rapidly depleting and contaminating the rivers, streams and aquifers of northern Alberta and the Northwest Territories, thereby contributing further to the unfolding ecological nightmare.

Ninety-three percent of the bitumen in northern Alberta is located too deep inside the Earth's crust to be mined. So, as we have seen, another method, in situ, is used, in which steam is injected deep inside the Earth in order to "melt" the bitumen so that it can be pumped to the surface. This process requires a great deal of fresh water. The generally accepted range is that it takes between two and five barrels of water to produce one barrel of crude oil from the tar sands. As a result, tar sands production puts a lot of stress on water sources in Alberta. The tar sands are in the prairie region of Alberta, an area that is subject to periodic (occasionally long-term) droughts.

The Alberta government provides the oil and gas industry with licences to draw the water they need for their operations. The petroleum sector alone is allocated 7 percent of all Alberta freshwater resources for its operations. In 2004, reports the Pembina Institute, three-quarters of the water used in conventional oil production came from a combination of both surface water (lakes and rivers) and fresh (non-saline) groundwater sources. Yet, unlike the water that is used by municipalities for households and commercial enterprises, most of the water used for oil and gas development does not return to the watershed from which it was taken. In other words, water used for conventional and unconventional oil and gas development is, for the most part, not renewable.

According to Pembina's 2006 report *Troubled Waters, Troubling Trends*,[225] the volume of water used for the in-situ operations in the tar sands has been growing at an alarming rate. Since 1999, the use of both fresh and saline water sources for in-situ drilling operations has increased five-fold. By 2004, the growth in tar sands extraction processes had become so rapid that the amount of fresh water used was three times higher than the Alberta government had forecast for that year based on 2001 data. The use of saline water was twice what had been predicted. The following year, 2005, the tar sands industry was allocated — for its mining and in-situ production — water withdrawals from the Athabasca River system that amounted to more than twice the volume of water used annually by the entire city of Calgary (population 1.2 million).

Alberta's largest river, the Athabasca, has become the prime source of water for the tar sands operations. From the Columbia Ice Fields near the Alberta–British Columbia border, the Athabasca flows down (across the northern part of Alberta) to its mouth in Lake Athabasca at the northeastern corner of the province, linking to the Peace and Birch Rivers along the way, as well as the mighty Mackenzie River, gateway to the far North. It is the third longest undammed river in North America.[226] Two-thirds of all the allocated water in the Athabasca River basin is designated for use by the tar sands industry. And a mere 10 percent of all the water taken from the Athabasca for tar sands mining operations is returned to the river. The rest of the water is diverted to the massive tailings ponds constructed by the companies to store the toxic refuse from their operations. In the winter months, when water flows are at their lowest, the tar sands operations can have serious impacts on the aquatic life in the river, particularly the fish populations.[227]

The huge water withdrawals from the Athabasca are not the only cause of the looming water crisis being generated by the tar sands. The Peace-Athabasca Delta, known as one of the world's largest freshwater deltas, is made up of a complex of wetlands and lakes. Through their strip-mining operations, the tar sands industry ends up destroying substantial areas of wetlands by removing and draining the muskeg that overlays the bitumen. In order to prevent the mine pits from flooding, the companies have also been draining the aquifer that underlies the bitumen. This, in turn, has negative effects on adjacent sources of groundwater as well as lakes and wetlands.[228] Here, groundwater withdrawals are the immediate concern. Like most other regions of the country, Alberta does not have a full and accurate accounting of all of its groundwater sources. Yet the tar sands industry is permitted by the provincial government to use a third or more of the fresh water from local groundwater sources for its in-situ operations. Without strict limits and rigorous monitoring of the industry's groundwater withdrawals, cumulative water takings are likely to become unsustainable. In fact, they may already have become so, simply because proper studies with benchmarks have not been carried out.

What's more, the volume of water takings will expand further as approved and planned tar sands projects come on stream to meet the five-fold increase in tar sands crude production projected for 2020. According to one study, the cumulative water takings by all of the existing tar sands operations (in 2005) from rivers, surface runoff and groundwater sources amounted to just over 150 million cubic metres.[229] When approved projects are added to existing projects, the total cumulative water takings are projected to rise to over 450 million cubic metres. And when the projects for tar sands crude production in the planning stage are added to those that are already approved or in operation, the cumulative water takings rise to around 675 million cubic metres. Nor will reductions in water use per barrel of oil produced due to new water-saving technologies offset this increase in the volume of water takings. No major breakthrough in water-saving technology for in-situ production is expected before 2030.[230]

It probably comes as a surprise to most people that so much water is required to produce crude oil. In fact, an energy–water nexus exists that is not all that well understood. We know, for example, that a great deal of water is required to produce hydro-electricity in this country. But, the water used for this is, for the most part, renewable: it can be used again. This is not the case with most of the water used for oil and gas production, especially tar sands crude. As we have seen, Alberta tar sands crude is proving to be among the most water-intensive hydrocarbons on the planet. Without sufficient supplies of water, the tar sands could not be mined. The industry understands this energy–water nexus. As Calgary oil consultant Bruce Peachy put it, "water availability will soon constrain future development."[231] He warned that it is unlikely that there will be sufficient water available to sustain production of three, let alone five, million barrels a day in the tar sands.

What's more, the availability of water will also be affected by climate change. David Schindler, who is perhaps Canada's foremost water scientist, has been studying the declines in the water flow of rivers on the Prairies to determine how these patterns are likely to be affected by global warming. Between 1971 and 2003, says Schindler, temperatures in the Athabasca region have risen by two degrees Celsius. As a consequence, the annual summer flows of the Athabasca River have declined 29 percent during this thirty-year period, while the river's contribution to other streams and rivers has dropped 50 percent. And there has been an analogous downward trend in the already lower winter flows of the Athabasca: they have dropped by an average of 1.5 metres per second. In an interview, Schindler calculated that this kind of decline in water flow is "equivalent to the demands of a new tar sands plant every two years."[232]

In effect, the work of Schindler and his colleagues shows that the energy–water nexus of rapid expansion of the Alberta tar sands puts it on a collision course with the effects of global warming. "If the climate continues to warm, runoff continues to decline and winter flows continue to decrease ... the water needs of the oil sands could reach a critical proportion of the winter low flow." They go on to warn: "Similarly, if the lower Athabasca River is affected by climate warming as projected for nine of its lowland tributaries, substantial declines in river flow may be expected between April and October as well." What happens in the Athabasca also affects the Peace River and the Slave River Delta in Great Slave Lake, which, in turn, is linked to the Mackenzie River system that flows all the way to the Beaufort Sea. Added to this is the rapid population growth in the tar sands production region that will further intensify pressure on the water supply.

The Alberta government, insists Schindler, has not yet accounted for climate change in its assessment of freshwater supplies, water licensing regulation, or water management plans for the tar sands region. Due to temperature spikes in this region, ranging between two and four degrees Celsius, over the past thirty years most of the Rocky Mountain glaciers have lost close to a third of their ice mass, thereby reducing the annual spring runoff into northern river systems. Nor has Alberta Environment sufficiently taken into account the rapid growth in in-situ production, which has resulted in water takings much greater than those forecast in 2004. As a result, Alberta Environment has not been able to accurately predict the increasing water demands of the industry for the period 2004 to 2020. Nor have adequate studies been done on the groundwater systems in the region to assess whether water capacities can meet the rapidly rising demand.

In February 2007, the Alberta government released a scathing report on water use by the tar sands industry stating that, "the Athabasca River may not have sufficient flows to meet the needs of all the planned mining operations." Written by former deputy minister of the environment Doug Radke, the report declared that the provincial government's ability to enforce environmental regulations is "inadequate." The report also stated that Alberta Environment had failed "to provide timely advice and direction to industry relative to water use." The report went on to describe the government's grasp of the cumulative water impacts of the tar sands as being "unclear, outdated and incomplete." A month later, Alberta Environment and Fisheries Canada announced a two-year interim plan designed to function as a stoplight for water withdrawals from the Athabasca — green light allows for full water withdrawals; yellow light indicates reductions in water withdrawals; and red light restricts allocations further. Yet, even when the red light signals drought conditions, the industry will still be allowed to withdraw enough water to fill fifty bathtubs per second.[233]

Moreover, in response to complaints by farmers along the North Saskatchewan River, the Radke report further admitted failure in adequately assessing the stress on water sources that has been, and will be, generated by the building of up to nine multi-billion-dollar upgrader facilities for the production of tar sands crude. If these nine upgraders are built as planned between now and 2020, they will require ten times the amount of water currently used by the city of Edmonton. All of this water is to be supplied by the North Saskatchewan River, which is only a third the size of the Athabasca. According to Alberta's new Water Management Framework for the North Saskatchewan, water withdrawals will be limited during low water flows; but they will not be stopped, even during critical low flow conditions.[234] An engineering study conducted for local governments estimates that water takings from the North Saskatchewan by the upgraders will amount to an increase of 278 percent by 2015, and 339 percent by 2025. For each upgrader facility, the daily water footprint is the equivalent of six to eight Olympic-size swimming pools (between sixteen and twenty million litres of water).[235] According to Schindler, the Alberta government's water framework assessment of the impacts on the North Saskatchewan is "pretty hollow," because it does not account for the river's declining water flows nor for the effects of climate change.

Yet the looming water crisis of the tar sands is due not only to the depletion of freshwater sources, but to its contamination. Approximately 90 percent of all the water used in the mining of the bitumen is stored as tailings in thick, soupy lake compounds or tailings ponds on both sides of the Athabasca River.[236] Together, these tailings ponds cover an area of fifty-five square kilometres; if they were drained into Lake Erie today the cumulative tailings would cover the lake bed to a depth of twenty centimetres. Given another decade, these tailings ponds will cover an area of 150 square kilometres, nearly three times their current size. For every barrel of oil produced, says the Alberta Chamber of Resources, six barrels of sand and one-and-a-half barrels of tailings are dumped into these ponds. These tailings contain salts, heavy metals, toxic hydrocarbons and pollutants such as napthenic acids and polycyclic aromatic hydrocarbons (PAHs).[237] This heavy concentration of toxins and other pollutants poses a direct threat to the fish and bird populations in the region.

The tar sands tailings compounds involve 10 dams, some of which reach as high as 100 metres. The largest one is the Syncrude Tailings Dam. Built in 1973, this Syncrude tailings pond covers an area of twenty-two square kilometres and holds some

540 million cubic metres of water, sand and tailings. Each day, Syncrude dumps 500,000 tons of tailings into its tailings pond. According to the US Department of the Interior, the Syncrude Tailings Dam has been, until recently, the world's largest dam. But, with the completion of the Three Gorges Dam in China in 2008, the Syncrude dam will be ranked the second largest in the world based on the volume of construction materials used. The toxin-laden muck that floats on top of the Syncrude and other tailings ponds poses a constant threat to ducks and other migratory birds who unknowingly land on the water surface, become soaked in oil, drown, or are poisoned. Syncrude and the other operators use scarecrows and propane cannons in an effort to keep birds from landing on their tailings ponds, which have not proven to be foolproof by any means.

Perhaps nothing more dramatically illustrates this problem than the sudden massacre of ducks that occurred in late April 2008. A flock of over 500 ducks landed on the Syncrude tailings pond, which is situated along a major migratory route for waterfowl. Only three of the ducks were rescued while more than 500 perished. Around the world, news of the ducks that died in Syncrude's tailings pond spread like wildfire. Images of dead ducks smothered in oil appeared on the front pages of newspapers and on television news broadcasts. Instantly, the incident sparked public furor and further tarnished Canada's environmental reputation. Although Syncrude officials apologized and Prime Minister Harper acknowledged the "terrible tragedy" that had occurred, the fact remains that the tailings ponds are highly toxic and pose an ongoing and profound environmental threat. The incident also proved to be a major setback to Premier Stelmach's 25-million-dollar campaign to assure the US and the rest of the world that the Alberta tar sands are "environmentally friendly."

The dangers of leakage from these tailings ponds into nearby groundwater systems are growing day by day. Randy Mikula, who works with Natural Resource Canada at the CANMET Energy Technology Centre in Devon, Alberta, and has been studying tailings waste matters for twenty-two years, calls the leakage of toxins into freshwater systems from the tar sands tailings ponds a "frightening" issue.[238] Every tailings pond contains toxins such as PAHs (polycyclic aromatic hydrocarbons) and napthenic acids, which are known killers of fish. Not only are leakages an apparently intractable problem but there are growing concerns about the prospects of a major dyke failure. As various engineering studies have concluded, waste-containment dykes have proved to be very unreliable. In 2003, the intergovernmental Mackenzie River Basin Board warned that a dyke failure at one of the tar sands tailings ponds "could have a catastrophic impact on the aquatic ecosystem of the Mackenzie River basin."

In a report entitled "Canada's Toxic Tar Sands," the eco-watchdog group Environmental Defence describes the toxic pollution being generated by the tar sands industry as "a slow motion oil spill in the region's river systems."[239] Citing specialists in the field, the report contends that the tar sands may be worse than the *Exxon Valdez* oil spill. The toxic chemical levels are already high and will only get worse as the tar sands boom continues, they say. Due to "notorious carcinogens in sediments and water ways," fish and game animals in the region "are found covered with tumours and mutations." Downstream, First Nations experience the damage first-hand: "There's deformed pickerel in Lake Athabasca ... Pushed in faces, bulging eyes, humped back, crooked tails ... never seen that [before]."[240] Fish frying in a pan "smells like burning plastic." According to Environmental Defence, one study by a company in the tar sands concluded that arsenic levels in moose meat from the region could be as much as 453 times the acceptable levels.

Of all the toxic chemicals generated by tar sands production, the ones of primary concern are naphthenic acids, mercury, arsenic salts, benzene and PAHs. Studies by independent scientists show that these chemicals constitute a toxic hazard to both humans and wildlife in the region. The sediment levels of PAHs in the Athabasca Delta, for example, are currently twice the threshold known to cause cancer in fish. Although the tar sands industry claims that various toxic chemicals occur naturally in this region, other independent studies show that the levels are steadily rising. In a study conducted for the Fort Chipewyan Community Heath Authority, Dr. Kevin Timoney compared current data on toxic chemicals with public findings recorded for these chemicals during the 1970s, 80s and 90s and discovered some alarming trend lines: mercury levels are now as much as 98 percent higher; dissolved arsenic levels have jumped by as much as 466 percent; sediment arsenic levels have risen by as much as 114 percent; and alkylated PAH levels have increased by as much as 72 percent.[241]

These and other toxic chemicals are concentrated and stored in the massive tailings ponds or, more accurately, toxic sludge reservoirs created by the tar sands industry. "These masses of toxic soup," says Environmental Defence, "have now grown so big that they can be seen by the naked eye from space." Huge pipes from each of the operating tar sands plants transport toxic sludge into these tailings reservoirs twenty-four hours a day, seven days a week. Like all tailings ponds elsewhere, says Environmental Defence, these toxic sludge reservoirs leak into nearby rivers and groundwater systems. In 1997, for example, Suncor actually admitted that its Tar Island tailings pond leaks almost 1,600 cubic metres of toxic liquid daily into the Athabasca River.

This is why Environmental Defence maintains there is a "giant slow motion oil spill" already taking place in the waterways of the boreal forest. In addition to the Athabasca, which is in the most immediate danger, the other major waterways affected are the Peace River, Slave River, Muskeg River, Slave Lake, Slave Delta and the Mackenzie River, which flows through the Mackenzie Delta to the Beaufort Sea. In 2003, the Mackenzie River Basin Board declared that "an accident related to the failure of one of the oil sands tailings ponds could have a catastrophic impact on the aquatic ecosystem of the Mackenzie River Basin due to the size of these ponds and their proximity to the Athabasca River."[242]

Downstream from the existing tar sands plants, it is First Nations communities that are most directly affected by the flow of toxic sludge. In particular, communities of the Mikisew Cree and the Athabascan Chipewyan have actively protested against the poisoning of their waterways and the health and environmental hazards facing their peoples. So, too, have the Dehcho First Nation of the Mackenzie Valley in the Northwest Territories, the Tlicho, Sahtu, Gwi'chin First Nations further north and the Inuvialuit of the Beaufort Sea region. The entire waterway of the Mackenzie, Canada's longest river (1,800 kilometres), is under dire threat, and its peoples, many of whom live traditional lives on the land, are fearful of what the future will bring.

Although the Alberta and Canadian governments have an obligation to oversee the health and environmental impacts of industrial megaprojects such as the tar sands, they have largely abdicated this responsibility to the industry, allowing it to regulate itself on a voluntary basis. When it comes to monitoring water pollution, for example, the Alberta government has outsourced much of this responsibility to the Regional Aquatics Monitoring Group (RAMP). Although RAMP includes representatives of local community groups, First Nations, and environmental organizations, as well as scientists, the tar sands industry plays a particularly active and dominant role in determining what RAMP does. It's a classic case of "the fox guarding the henhouse," says Environmental Defence. RAMP's testing of the sediment in the Athabasca River bed is erratic at best. According to independent scientists, RAMP's monitoring and testing programs are highly questionable. "Changes in methods and means of reporting," for example, says Dr. Timoney, "undermine the utility of results,"[243] while a specialist in the toxic impacts of oil on aquatic ecosystems, Dr. Peter Hodson of Queen's University, Kingston, Ontario, maintains, "you'd be hard pressed to say the current monitoring effort is sufficient to answer the questions that have been raised."[244] Both the Mikisew Cree and the Athabasca Chipewyan have withdrawn from RAMP because of its failure to adequately monitor the water.

This looming water crisis has national impacts as well. The rush to develop the tar sands poses a threat to Canada's water security in the long run. The Athabasca–Peace–Mackenzie watershed, where rivers and lakes are interconnected with one another, is one of the largest freshwater watersheds in the world. It is estimated that one-sixth of Canada's fresh water drains into it, so what happens in this enormous freshwater system affects the water security of Canada as a whole. The House of Commons Natural Resource Committee acknowledged as much in its 2007 report, recognizing that the water crisis being generated by rapid tar sands development posed a serious problem. But for the more than 360,000 First Nation peoples whose ancestors have inhabited this land since time immemorial, access to clean, fresh water from this massive watershed is a matter of life and death.

Boreal Destruction

Reminiscent of the strip mining of the Appalachian mountain tops for coal, the tar sands industry is allowed to rip up the boreal forests of northern Alberta to produce crude oil from bitumen located closer to the earth's surface. Roughly half of current tar sands production comes from open-pit mining, as distinct from the in-situ processes, and the damage being done to the boreal forests is extensive. When the tar sands sites in the Athabasca, Peace and Cold Lake regions are in full operation, a sizeable portion of the boreal land mass will have been transformed into a moonscape. Viewed from outer space, this moonscape appears as a huge crater in the forest. In many ways, this moonscape in the midst of the boreal landscape looks remarkably similar to the Appalachian mountains that have had their tops ripped off to get at the remaining coal reserves. The situation is also reminiscent of the ecological damage caused by massive mining operations in places such as Elliot Lake and Uranium City, and the clear-cut logging of British Columbia.

To mine the shallow portions of the tar sands, some of the most advanced mining machinery and technology is used. After the timber has been stripped from the landscape and the protective muskeg ripped off the earth, huge hydraulic shovels dig open mine pits of up to five kilometres in diameter.[245] One example of these shovels is the 495HF Bucyrus, which is made in Wisconsin and costs over 15 million dollars. The Bucyrus deposits the massive chunks of bitumen, minerals and soil into huge, forty-tonne trucks. These are the Caterpillar 797B trucks, made in Illinois, and are no less than a storey-and-a-half high. The giant caterpillars transport the material to the company plants where the bitumen is separated and the rest, as well as most of the water used in the process, is discarded as waste in the massive, toxic tailings ponds.

The forests of northern Alberta are part of the vast Canadian boreal landscape that stretches across the northern tips of the provinces and the southern edge of the northern territories. Canada's boreal forests contain about 25 percent of the remaining intact forests in the world. An "intact forest landscape," say the Rainforest Action Network and ForestEthics, "include[s] forests and an abundant variety of natural ecosystems, such as wetlands, mountains and tundra."[246] The Canadian boreal forests are often referred to as the northern lungs of the planet, complementing the Amazon Basin forests of Latin America, the earth's southern lungs. As an intact forest system, the boreal provides an uninterrupted habitat for animals sensitive to human incursions, including caribou, moose, bears, wolves and an entire system of animal life, as well as breeding grounds for a rich diversity of waterfowl and other migratory birds. So far approximately two-thirds of the Canadian boreal landscape remain undisturbed by industrial development.

In an age of climate warming caused by carbon emissions, Canada's boreal forests become a matter of prime importance. Globally speaking, the boreal is the largest land reservoir of carbon on earth. Approximately 22 percent of the total carbon stored on the land surface of the planet is already housed in the boreal forests. Due to cooler temperatures in boreal climates, decomposition is slow, resulting in deep deposits of organic soils that are thousands of years old. Of the world's remaining forests, 50 percent are in the boreal range stretching across Canada, Alaska, Russia and Scandinavia. According to scientists, it is estimated that the Canadian boreal forest and peatland ecosystems store 180 billion tonnes of carbon, the equivalent of twenty-seven years of carbon emissions from burning fossil fuels on the planet as a whole.

In December 2007, at the United Nations' Conference on Climate Change in Bali, the International Boreal Conservation Campaign (IBCC), in conjunction with four other organizations, released a set of maps portraying Canada's boreal forest as "the world's largest terrestrial carbon storehouse." As Jeff Wells, the senior scientist for the IBCC, put it: "The Boreal Forest is to carbon what Fort Knox is to gold." The maps were put together by Global Forest Watch Canada based on a detailed analysis of extensive data available on the region. The three maps the IBCC presented showed the distribution of peatlands, the distribution of permafrost, and the levels of organic carbon in soils across Canada's boreal forest; together, these three features of Canada's boreal are key for carbon storage in the future.

The first carbon storage feature of Canada's boreal is its vast peatland areas, the largest to be found anywhere in the world. Having the capacity to store six times as much carbon per hectare as forested mineral soils, peatlands are now recognized as an extremely important storehouse for carbon. Globally speaking, peatlands account for between 20 and 25 percent of all soil carbon stores. As the IBCC's peatland map shows, an estimated 12 percent of Canada's land mass, or 1.14 million square kilometres, is composed of peatlands, making this country a world leader in peatland carbon storage. Canada's vast boreal peatlands stretch from Labrador and Quebec in the east to beyond the Mackenzie Valley in the west. The most significant peatland regions are concentrated in northern Ontario and Manitoba.

The second carbon storage feature of Canada's boreal region is its huge areas of permafrost. Since permafrost is permanently frozen ground, it has a much lower decomposition rate, therefore providing a lasting natural storehouse for the earth's carbon. Close to 50 percent of Canada's total land mass is composed of permafrost. As the permafrost map released by the IBCC in Bali shows, the northern portions of Canada's boreal, especially in the western region of the country, are laden with vast areas of carbon-rich permafrost. Permafrost covers roughly a third of the boreal region. As water scientist and ecologist David Schindler was quoted in the IBCC's press release for the Bali conference, "The carbon frozen into Canada's permafrost … is one of North America's largest stores of carbon. It's similar to a bank vault containing one of the world's most valuable and most influential resources for impacting climate change."

The third carbon storage feature of Canada's boreal region is its soils. Close to 90 percent of the organic carbon found in soils across Canada is located in the ecosystems of the boreal forest and the tundra. The colder soils of the boreal are capable of storing and retaining much more carbon than the soils in the more southern, warmer climates in the rest of the country. According to the soils map released in Bali, there are several key areas where the boreal soils contain significant stores of carbon, even though soils generally have less capacity for carbon storage than either peatlands or permafrost. But the boreal soils depend on intact forests. Any major deforestation would have serious negative impacts on them. Although the federal government has announced plans for new protected areas in the Northwest Territories, and the Ontario government has pledged to work with northern and Native communities to protect the boreal forests and land, only 10 percent of Canada's boreal region is legally protected from major industrial developments such as the tar sands.

The value of Canada's boreal forests to the future of North America and the planet as a whole in an age of climate warming is perhaps incalculable. Nevertheless, the Canadian Boreal Initiative (CBI) commissioned a study to determine whether a dollar figure could be put on the wealth of Canada's boreal. Entitled "The Real Wealth of the Mackenzie Region," the CBI report attempts to assess the natural capital value of the Mackenzie Valley as a northern ecosystem of Canada's boreal.[247] According to the report's findings, the Mackenzie region stores sixty-seven billion tonnes of carbon, the equivalent of 300 years of Canada's carbon emissions based on levels in the year 2000. The net value of the Mackenzie's boreal carbon bank, says the report, is around 3.7 trillion dollars.[248] The value of all benefits from the Mackenzie boreal system is estimated to be 93.2 billion dollars per year, two-and-a-half times the net value that would be realized from the extraction of the natural resources of the region.[249]

The value of Canada's boreal as a carbon sink, however, could dramatically change as a result of climate warming trends. An estimated 60 percent of Canada's peatlands are expected to be seriously impacted by climate changes, notably in the boreal and subarctic regions.[250] Instead of retaining their capacity to store carbon, the boreal regions of Canada could become major emitters of carbon. The drier conditions and lowering of water tables predicted for the north, for example, means that previously flooded peatlands will be exposed to the air, leading to faster decomposition of the organic materials of which they are made, and hence the emission of the stored carbon.[251] Moreover, the exposed peat will be more susceptible to fire, which could result in even more carbon being emitted. In the same way, the melting of permafrost will lead to more decomposition of peat, again causing a rapid increase in the emission of carbon dioxide, methane and other greenhouse gases.

Meanwhile, the growing number of tar sands operations poses an immediate threat to Canada's boreal carbon sink. Take, for example, the Fort Hills Project of Petro-Canada, Teck Cominco and UTS Energy located in the Athabasca River corridor and the McClelland Lake Wetland Complex. Fort Hills is a strip-mining operation that threatens what environmentalists call a "world-class wetland habitat," home to a rich diversity of wildlife, including endangered species, and rare plant species. Called the "Jewel of the Boreal," the McClelland Lake Wetland Complex includes McClelland Lake, twelve sinkholes and two ancient patterned fens that provide groundwater recharging and habitat for plant and animal species.[252] The Canadian Parks and Wilderness Society has called for the protection of the McClelland Complex and the Athabasca River corridor from all industrial development. Regardless of the damage it will do to the boreal, the three companies are proceeding with the Fort Hills Project. Petro-Canada has even argued against a broader federal environmental review of the project.[253]

It is not only the strip-mining operations of the tar sands such as the Fort Hills Project that pose a direct threat to the future of Canada's boreal. The tar sands in-situ processes are also a serious threat. As we have seen, the already large and rapidly growing volumes of water required for the in-situ process to bring the bitumen to the surface ends up draining water from the region's river systems and aquifers. The peatland areas of the boreal depend on these freshwater sources in order to function as carbon sinks. At the same time, both in-situ and strip-mining operations in the tar sands further threaten Canada's boreal with their own greenhouse gas emissions: the burning of natural gas in the process to separate the bitumen from the sands is most certainly contributing to the acceleration of the climate warming that poses a threat to Canada's boreal as a natural carbon storehouse for the planet.

In effect, Canada's boreal forest is under siege in Alberta because of the major rush to develop the tar sands to provide crude oil to the US. Although there is increasing public concern in Alberta about the environmental damage being done, most people in the province are simply unaware of the growing number of oil companies that are claiming rights to tar sands territories. Nor is there much awareness of how the pace, scale and intensity of development in the tar sands region is being managed, or mismanaged. Between 1991 and 2005, reports the Pembina Institute, Alberta Energy leased 28,129 square kilometres of tar sands lands to oil companies, but in 2006 alone, leases were handed out for a whopping 15,424 square kilometres.[254] Once the oil companies secure these land rights, they start cutting access roads, setting up seismic lines and clearing land for exploratory drilling and open-pit mining.

According to Pembina, the tar sands mines currently in operation cover only 420 square kilometres.[255] However, 1,000 more square kilometres have already been approved by Alberta Energy for mining. Another 3,000 square kilometres, an area almost five times the size of Toronto, has been leased for strip-mining operations in the future. On top of this, an additional 3,000 square kilometres of the boreal forest will be cleared to make room for in-situ extraction of bitumen from the deeper tar sands deposits. These operations will require the construction of some 30,000 kilometres of roads in the region. In short, it is this legal process of land rights tenure that allows the oil companies in the tar sands industry to occupy and rip up the boreal forests of Alberta, and it has its own built-in imperative for growth and expansion.

The land tenure process that Alberta Energy has developed for the oil companies in the tar sands contradicts the principles and commitments to sustainable development made by the Alberta government and its Department of the Environment. In 1999, notes Pembina, the Alberta government came out with its Commitment to Sustainable Resource and Environmental Management (SREM) in which it declared that "environmental decisions will take into account economic impacts and economic decisions will reflect environmental impacts."[256] In 2005, the SREM program called for "using a strategic, systems approach [to resource development] driven by clear, concrete, agreed-upon outcomes and based on a sound understanding of our resources and environment." These principles, emphasized SREM, should be adopted across ministries in the Alberta government. SREM also called for "an effective management system" in which citizens, business, communities and governments would work together and take joint responsibility.

In 1999, the Alberta government also set up the Regional Sustainable Development Strategy to develop policy frameworks for environmental management in the tar sands region. A year later, the Cumulative Environmental Management Association (CEMA), a multi-stakeholder body, was created specifically to identify the ecological limits of the boreal forest and to assess the cumulative environmental impacts of the tar sands industry on the landscape and its wildlife. CEMA was expected to come up with a framework for managing the cumulative environmental impacts of the tar sands operations.[257] Initially due in 2002, the CEMA report still had not been completed by the spring of 2008, mainly because the tar sands companies who sit on the committee have a veto.[258] Meanwhile, Alberta Energy continues to grant leases, provide licences and approve projects for both strip mining and in situ without an environmental management plan that takes seriously the multiple impacts these have on the boreal region. And it does so without a land-use plan to guide its decisions.

However, coming up with a cumulative environmental management plan and a comprehensive land-use plan for the tar sands region will make little difference to the future of the boreal unless and until there are fundamental changes in the policies and operations of Alberta Energy. After all, Alberta Energy is the power centre and prime revenue generator of the Alberta government. In effect, warns the Canadian Institute of Resources Law: "Alberta Energy has played the spoiler on land-use and stewardship initiatives in the past. As the government's cash cow, it has the power in Cabinet to pursue a single-minded growth strategy for the oil and gas sector, with scant regard to cumulative environment effects."[259] When it comes to the future of the boreal and its role as a carbon storehouse for the continent and the planet, Alberta Energy may well appear to be the big spoiler. But it should be kept in mind that the driving force behind Alberta Energy is Big Oil itself and the US government.

The oil companies and the Alberta government insist, of course, that they have no intention of destroying the boreal. Quite the contrary, the corporations compete with each other for the reputation of "greenest" in the tar sands. Syncrude, for example, boasts about how it has reclaimed an old open-pit mine by laying sod, planting trees and installing a small herd of bison in a fenced-in area, along with a viewing area from which tourists can take pictures. Not to be outdone, Suncor shows visitors the narrow wooded corridor it has preserved on one of its old mine sites for animals to cross, complete with a map displaying sightings of moose, deer, fox and wolves. It all seems surreal, observes journalist and activist Hugh McCullum, like a southern city park or zoo stuck in the middle of the boreal forest.[260]

For its part, the Alberta government does issue guidelines for reclamation to the oil companies operating in the tar sands, requiring them to restore the landscape they have disrupted to its original condition. Yet, after more than forty years of tar sands development, only a small parcel of Syncrude's land lease for strip mining has been certified as reclaimed, under Alberta government guidelines. Indeed, says the Pembina Institute, Alberta Energy does not even regulate the reclamation rates of the companies under its approval process. Instead, the companies are free to set their own reclamation schedules and carry them out voluntarily. After 40 years in operation, Suncor has occupied and disturbed over 13,000 hectares of boreal lands and reclaimed, by their own standards, only around 950 hectares, none of which has been certified as meeting the quite liberal guidelines for reclamation set by the Alberta government. What's more, the goal of actually restoring this land to its original condition is probably unobtainable. As Lee Foote, a wetland specialist at the University of Alberta puts it, "no one really knows yet how to reclaim a fen, bog or peat land in the oil sands."[261]

Finally, new research is raising disturbing concerns that the warming of the north through greenhouse gas emissions may be triggering a self-perpetuating climate time bomb trapped in the permafrost. According to results of a 2008 study published by *Nature*, the thawing of the permafrost releases greenhouse gases into the atmosphere at a rate that is five times faster than anticipated. As thawed permafrost releases greenhouse gases, which, in turn, trap more heat in the atmosphere, more permafrost thaws, creating a continuous self-perpetuating cycle. "The higher the temperature gets, the more permafrost we melt, the more tendency it is to become a more vicious cycle," said Chris Field, director of global ecology at the Carnegie Institution in Washington, DC. In the summer of 2008, the journal *Science* published a study showing that one kind of carbon trapped in the permafrost, called vedoma, is much more prevalent than originally thought and may amount to 100 times the volume of carbon released into the air each year by the burning of fossil fuels.

When all is said and done, the environmental implications of the development of the Alberta tar sands, at present let alone in the near future, are simply staggering. The multiple impacts of massive greenhouse gas emissions on global warming, the rapid depletion and contamination of our northern water systems, and the destruction of a significant portion of the boreal, the world's natural carbon sink, will have devastating consequences not only for this country and continent, but for the planet at large. As Environmental Defence puts it, the tar sands enterprise is positioned to become "the most destructive project on earth." This is the ecological nightmare that's emerging and it's unfolding right here on our own doorstep.

Suncor mining operations in the tar sands. Courtesy David Dodge, Canadian Parks and Wilderness Society.

CHAPTER 6

Social Upheaval

The Alberta tar sands are not only proving to be an ecological nightmare. They are also a caldron of social upheaval. On the frontlines of the tar sands plants, social conflicts and tensions are brewing in local communities and amongst Aboriginal peoples, workers and farmers. For Albertans themselves, the spoils of the tar sands boom have by no means been equally shared. In turn, these social tensions and conflicts spill over into the rest of the country. What's more, the heavy concentration on the production of tar sands crude for export to the US is beginning to raise anxieties in the rest of Canada about our own energy security as a country. Added to this are the ongoing concerns about the use of Alberta tar sands crude for US military purposes and the potential for the exercise of US national security interests over oil and other vital resources in this country, all of which are ingredients for social upheaval.

Social Conflicts

You would think that in Fort McMurray, the hub of the tar sands boom, people would be humming all the way to the bank. After all, with several hundred billion barrels of oil in the region and a rapidly growing industry generating the highest paying jobs to be found anywhere on the continent, this boom could last as long as eighty to a hundred years. Here, tar sands workers eat at restaurants called "Fuel," drink at bars called "Oil Can," and gamble their wages away at casinos called "Boom Town."[262] But, while McMurray is awash in cash, or McMoney, as it's now often called, it is also beleaguered by a litany of social problems.

Fort McMurray's population is growing at a rate of 8 percent per year, reaching more than 80,000 in 2008, making it the province's third largest city after Calgary and Edmonton, but this northern Alberta frontier town only has the infrastructure to handle a population of little more than 10,000.[263] Basic municipal services such as garbage pickup, policing and fire protection are grossly overloaded. After the spring thaw the bare streets are littered with garbage, the schools are overcrowded and there is a pressing need for a new water treatment facility. Land for new housing is at a premium, despite the abundance of adjacent Crown land, and housing prices have been going through the roof. And then there's the narrow Highway 63, better known as the "highway to hell," used by workers fleeing to Edmonton on weekends, which is in desperate need of repair and upgrading.

The feisty mayor of Fort McMurray, Melissa Blake, has been lobbying the Alberta government relentlessly for increased funding in order to put in place the infrastructure and services needed to keep pace with the rapid growth. In May 2005, she and her fellow councillors petitioned the Alberta government, through the Municipality of Wood Buffalo, in which Fort McMurray is located, for a grant of 1.2 billion dollars.[264] The money would be spent over a four-year period: 353 million on roads and water, waste-

water, and recreation facilities; 236 million on primary, secondary and post-secondary education; 500 million on highway repair
and development projects; and 126 million on health and affordable (low-income) housing. In its initial response, the Klein
government only committed to 60 million dollars. But when
Mayor Blake threatened to use what municipal powers she had to
slow down development in the tar sands, another 530 million
dollars from provincial coffers was added, with the warning that
there must be no measures taken to impede development.

Yet, even with the province coughing up funds for municipal
services, the town faces other woes. The local economy is badly
skewed. On average, annual household incomes hover between
90 and 100 thousand dollars. Plumbers, electricians and pipe-
fitters, for example, earn well over 100,000 dollars a year working
for the big oil companies or their contractors, and their average
age is around thirty.[265] Not suprisingly, the municipal government and many local businesses and services can't compete when
it comes to hiring workers. Never mind the difficulties the service industries have finding and keeping workers in this climate,
local schools and hospitals are finding it very difficult to keep
good teachers and medical staff because they cannot afford to pay
competitive wages, which compounds the problem of providing
public services to the growing population. Moreover, without
adequate wages, almost no one can afford to live in Fort McMurray, where housing costs have become a nightmare for renters
and first-time homebuyers.

The town's housing crisis is perhaps the most pressing problem. It's not uncommon to find eight or ten people sharing an
apartment; others sleep in their trucks. Workers pitch tents along
the Athabasca River, or rent garages or trailers. Hundreds of
homeless men and women, many of them on drugs, "walk the
streets like zombies."[266] Under these conditions, the crime rate
has soared. Assaults in Fort McMurray are reportedly 89 percent

higher than in the rest of Alberta. Arrests for drug-related offences are 215 percent higher, and arrests for impaired driving 117 percent higher than those recorded elsewhere in the province. Every week, an estimated 6.5 million dollars' worth of crack cocaine makes its way up the highway to hell. The town's detox centres are overloaded, suicide counselling is all but maxed out and there is a crying need for more doctors.

The distribution of wealth from the tar sands oil boom throughout Alberta is similarly skewed. According to an Environics poll conducted in March 2007, half of Albertans either felt worse off (17 percent) or about the same (34 percent) as a result of the boom.[267] The Parkland Institute found these poll results surprising given the incredible wealth generated by the boom, and decided to probe further. In the June 2007 Parkland report, *The Spoils of the Boom*, Diana Gibson found that "the benefits of the boom are disproportionately going to high income Albertans, most notably the top income bracket."[268] In other words, "Alberta's wealth is trickling up not down." Although middle-income Albertans have seen small increases in their incomes, this is not really due to the boom but because they have been working more hours to maintain their standard of living. Moreover, housing prices and rents have been skyrocketing. Between January 2006 and June 2007 new home prices increased by 65 percent in Calgary, while in Edmonton the average selling price for a condo or single family home jumped 52 percent.

At the same time, the Parkland study shows that many low income Albertans "are worse off" because of this boom. Since 1996, there has been a 458 percent growth in the number of homeless people on the streets of Calgary.[269] In 2006, when the boom was in full swing, homelessness in Edmonton grew by 19 percent. Despite record levels of wealth being generated, Alberta still has the lowest minimum wage of any province in the country. While there was a minimum wage increase in 2005, it barely

covered losses incurred in recent years because of inflation, and
was nowhere near sufficient for a booming economy. Because
Alberta's social assistance rates are not indexed to inflation, families and individuals who rely on this have seen their real incomes
drop. In Alberta, social assistance for a single parent family is a
mere 12,326 dollars annually, which, says the National Council
on Welfare, is only 48 percent of the annual income needed just
to reach the poverty line. As rising rents and inflation eat away at
their limited incomes, the Alberta boom is a bust for this sector
of the population.[270]

The real benefits of the Alberta oil boom, however, are not only
going to the highest income Albertans but to the rapidly growing
revenues and profits of corporations in the petroleum sector,
most of which are foreign-owned. In 2005, reports Statistics
Canada, a record high was achieved in corporate operating profits, led by the oil and gas sector. When it comes to the oil and gas
sector in this country, says Statistics Canada, nearly half the assets
and over half the revenues are foreign-owned. As a result, it is foreign shareholders that are benefiting most from the boom. The
Parkland study looked at ten oil and gas companies in Alberta
and found that their revenues in 2005 had risen on average 26.5
percent over 2004. During the same period, dividends amounted
to 213 million dollars, an increase of 28.6 percent over 2004.[271] In
short, the Alberta boom has been translating into rising corporate profits, and the main beneficiaries are foreign shareholders.

Underlying this is a conflict over the royalty payments made by
the oil industry for the extraction and production of the resource.
Since it is the people of Alberta who, constitutionally speaking,
own the oil in the tar sands, the oil corporations operating there
are obligated to pay them royalty fees for the extraction, and
depletion, of this natural capital. However, in 1996, both the federal and Alberta governments agreed to a set of recommendations
put forward by an industry-led task force that virtually granted

the tar sands companies a long royalty-payment holiday. The agreement, signed by all three parties (i.e., the two levels of government and the industry) amounted to a royalty fee of only 1 percent on the revenues of all tar sands ventures until their capital costs are recovered. When then premier Ralph Klein and then prime minister Jean Chrétien made the announcement, Chrétien described the lucrative tax break for the industry: companies operating in the tar sands could write off 100 percent of their capital costs, including cost overruns. At the time, critics condemned the deal as a theft of the natural birthright of Albertans.

In short, both Ottawa and Edmonton had decided to subsidize the tar sands industry through a generous incentive package comprised of federal tax writeoffs and low provincial royalty rates. Alberta's low royalty fees had not changed since they were established back in the mid-1980s when oil prices were just over 10 dollars a barrel. After the new royalty regime deal was struck in 1996, the Alberta Heritage Fund was unable to keep pace with its counterparts in other oil-producing jurisdictions. According to a study by the Parkland Institute, the Alberta Heritage Fund had 13.7 billion dollars in 1996, compared with the heritage funds of Norway at 11.8 billion and Alaska at 26.5 billion.[272] By 2002, even though tar sands production had grown by an estimated 74 percent since 1996, Alberta's fund had dropped to 11.8 billion dollars. During the same period, Norway's fund jumped to 101 billion dollars while Alaska's increased modestly to nearly 36 billion. Regardless of the booming tar sands, says Parkland, the Klein government took in less than half the revenue per unit of oil that the Lougheed government did in the mid-1970s from conventional oil production.

If the Alberta government had applied adequate royalty rates for petroleum production to the tar sands industry, it would have had an abundance of resources with which to resolve Fort McMurray's municipal services crunch plus the social inequities and environmental challenges facing the province. But instead Alberta's royalty regime went from bad to worse. Shortly after taking office as premier in 2007, Ed Stelmach appointed an independent Royalty Review Panel to conduct hearings throughout the province and make recommendations. When the panel of experts submitted their recommendations in October 2007, calling for a modest increase of 22 percent in provincial royalty rates for Albertans, the Parkland Institute said the panel was "Selling Albertans Short."273 But, Premier Stelmach rejected every one of the panel's proposals. Instead, he put forward his New Royalty Framework calling for a mere 2 percent increase in Albertan's share of oil and gas revenues. The difference between the premier's and the panel's proposals, calculated the Pembina Institute, is "4.8 billion dollars in revenues over the life of a typical in-situ project and 11.6 billion dollars over the life of a typical mining project."274 What's more, the 2008 Organization for Economic Co-operation and Development's *Economic Survey of Canada* appears to confirm much of Parkland's and Pembina's critique of Alberta's revised royalty regime, and encourages a more Norway-like investment fund, whereby windfall returns can be spread over a long period of time so as to prevent the overheating of the economy and related negative effects.275

Meanwhile, the tar sands have caused an outbreak of social conflicts brewing amongst different constituencies, including workers, farmers and First Nation peoples. In the pro-business climate of Alberta, labour unions have to constantly watch their backs. Since the tar sands industry generates thousands of high-paying jobs for skilled and semi-skilled workers, the companies are always looking for ways to cut their costs by keeping wages down. One way has been to hire non-union workers and/or cheap foreign labour. In an unprecedented move, the Klein government enabled one of the major tars sands companies, Canadian Natural Resources Ltd., to employ non-union workers for its Horizon Project in 2005. Using an obscure clause in the Alberta Labour Relations Code, the cabinet granted CNRL "special status" to use non-union workers and foreign labour at cheap wages for both the building and operation of its new Horizon plant.[276] It also allowed the company to re-write the rules governing the employment of tradespeople for the construction of similar tar sands plants in the future, including those on wages, overtime and travel.

Tar sands companies such as Suncor continuously argue that there is a shortage of qualified and skilled workers. But the Alberta Building Trades Council (ABTC) counters these claims by pointing to concrete cases such as the construction of Suncor's Athabasca Oil Sands plant, completed in 2003, that used up to 15,000 workers supplied by building trades unions in the province. If Alberta workers were taken up with other construction projects, says the ABTC, then the labour shortfall could be made up through other union affiliates across the country from a nationwide pool of more than 400,000 skilled workers.[277] The Alberta Federation of Labour, however, sees the worker shortage issue as a ruse to give the government an excuse to permit the companies to cut their costs by employing non-union workers or cheaper labour from other countries. The government's "special status" order, says the AFL, sets a dangerous precedent: hiring less qualified and less experienced tradespeople runs the risk of lower work quality and more accidents. It also provokes job-site conflicts between unionized and non-unionized workers.

These labour conflicts have intensified due to the involvement of organizations such as the Christian Labour Association of Canada (CLAC). Tar sands companies, such as CNRL, like to deal with CLAC because it offers lower wage scales and is known for striking "sweetheart" deals with their employers. Recognized in only five provinces, including Alberta, CLAC is seen by organized labour as "as a handmaiden of management and a K-Mart union."[278] Once CLAC strikes a deal with a company, it sets the standards for all other workers in the company and has an impact on other companies in the industry. Many ABTC workers will not work on a CLAC site, which led CNRL to set up separate camps for its workers to avoid the outbreak of hostilities between the two groups. For its part, Suncor proposed that this kind of acrimony may have to be resolved by importing labour from other countries.

Although Suncor and CNRL still claim they will not hire workers from other countries, many of their contractors do. One Suncor contractor, the Ledcor Group of Vancouver, received Ottawa's approval to import 700 skilled oil workers from Latin America. And now, under the federal government's Temporary Foreign Worker Program, over 25,000 workers have arrived from dozens of countries including the Philippines, India, China and Poland. But the AFL contends that besides the problem that importing cheap labour from other countries undercuts workers' rights in Canada, the foreign workers are often mistreated when they get here. In a report released in December 2007, the AFL documented cases in which foreign workers were paid lower wages than promised, lived in crowded or squalid housing, were forced to pay hundreds or even thousands of dollars in illegal fees, and faced threats of deportation when they complained about the treatment.[279]

Elsewhere, farmers and rural communities are increasingly upset by the rapid expansion of "Upgrader Alley." Located in Alberta's industrial heartland, a 32,000-hectare area about a half-hour's drive northeast of Edmonton, a corridor of upgrader facilities is being built along the North Saskatchewan River to upgrade the bitumen from the tar sands plants into synthetic crude oil. As noted in Chapter 3, Shell Oil and Dow Chemical have already built upgrader plants in the region. Two more are currently under construction, seven are going through the regulatory process and one other is planned but not yet announced.[280] Just a few decades ago, this was all farmland. While a significant number of farm families remain in the area, their lives have been dramatically disrupted by the rapid invasion of dozens of refineries, petrochemical plants and now the tar sands upgraders. Rampant air pollution, in particular, has become a major concern. People living near Shell's upgrader plant, for example, are exposed to routine emissions of sulphur dioxide and other toxic chemicals, as well as a heavy dose of greenhouse gases caused by the burning of natural gas in the upgrading process. Since 2003, when Shell's first upgrader went into operation, there have been on average four to five accidents a year, mostly gas leaks. One farm reported having lost 45 dairy cows out of a herd of 140 over a two-year period.[281]

Another flashpoint for farmers is Upgrader Alley's intensive use of water. The upgrader plants use huge quantities of water in the process of upgrading the bitumen into crude oil. This water is taken from the North Saskatchewan River. Up and down the North Saskatchewan, however, farmers are complaining that the river levels are getting lower and lower every year. In his own studies of the changing water levels of Alberta's rivers, David Schindler confirms that the flows of the North Saskatchewan have been in decline. The new upgrader plants, says Schindler, are accelerating the process. What's more, the Alberta government is committed to expanding upgrader capacity to the point where 80 percent of the bitumen from the tar sands will be upgraded by these facilities before being exported south to US refineries. Yet, no government studies have been undertaken to see if water supply can meet water demand. As we saw in the last chapter, the 2007 Radke report, *Investing In Our Future*, released by the Alberta government, admitted as much, saying: "Alberta Environment has not had the opportunity or the resources to undertake a review to determine whether there is sufficient water available from the North Saskatchewan River to support these upgraders."[282]

For the farmers and rural communities living near Fort Saskatchewan, there are parallels to be drawn with what is happening to Fort McMurray. Just as the rapid expansion of the tar sands industry has overwhelmed Fort McMurray to the point that people are saying the growth is "too big" and "too fast," so too the residents of Fort Saskatchewan and surrounding towns are worrying about the upgrader boom that is about to hit them. The huge demands on municipal services experienced by Fort McMurray — which result in overcrowded schools, overloaded hospital services, limited water treatment capacities, inadequate police and fire protection, poor road conditions, strained garbage collection services and full land-fill sites — weigh heavily on the minds of people in Fort Saskatchewan, neighbouring towns and Edmonton itself.

Meanwhile, north of Fort McMurray another kind of social conflict generated by the tar sands has been brewing for more than two decades. In 1982, the residents of Fort Chipewyan and Fort McKay noticed that the fish from the Athabasca River tasted like gasoline. It turned out that Suncor had had an oil spill, which the government neglected to act on, and an out-of-court settlement was reached between the company and a group of fishermen. Earlier, Syncrude had also had an oil spill. As a result, the Aboriginal peoples who inhabit the communities downstream from the big tar sands operations found that their livelihood, which depends on fish and the Athabasca River, was being threatened. Later, community alarm bells rang over health concerns as the number of cancer cases and other diseases unheard of before in the region rose dramatically. After serving as the visiting family physician since 1993 in Fort Chip (as it is known locally), Dr. John O'Connor started to speak out publicly in 2003 about these emerging health concerns in the community.[283]

Among his patients, O'Connor diagnosed an alarming number of rare cancer cases and autoimmune diseases such as rheumatoid arthritis and lupus. The incidents were too numerous to ignore given the small population. After conducting a battery of blood tests and sending them to an endocrinologist in Edmonton for further examination, O'Connor began to search for the causes. Autoimmune diseases, for example, can be caused by arsenic, benzene and PAHs, which, as we have seen, are all present in the tar sands. O'Connor found that the sediment in which the bitumen is located also contains significant quantities of arsenic. During in-situ drilling operations, arsenic is brought to the surface where it is mixed with surface water. As a result, high concentrations of arsenic end up in the company tailings ponds, which, in turn, leak into the surrounding groundwater systems. At the same time, the Alberta government allows the companies to dump sixty kilograms of arsenic a year into the Athabasca River.[284]

O'Connor tried to get a baseline study of the health of the people living near the tar sands conducted, something that had never been undertaken. When Shell Oil and CNRL applied for licences to build two new strip mines, O'Connor and the chief of medicine at the hospital in Fort McMurray, Dr. Michael Sauve, testified about the heath concerns in Fort Chip, and the need for such a health study. In granting Shell and CNRL their licences, the Alberta Energy and Utilities Board required the companies to fund a baseline health study of the Fort Chip community. But, as of 2008, that study has not been undertaken. Health Canada was also lobbied by O'Connor to do the study but ignored him until the issue received more public attention. Suddenly, in April 2006, Health Canada and the Alberta Health and Wellness department announced that they would do a study of diseases at Fort Chip. In July 2006, the health investigators announced that their study of the town's medical records did not reveal sufficient evidence of increased cancer rates or abnormal rates of other diseases to establish the claim that they were being affected by the tar sands operations. When questioned by members of the Aboriginal community, the health investigators admitted there had been a slew of omissions. Their study failed to test for chemicals in the peoples' bodies, their food or the river itself. Nor had they examined the medical records that O'Connor had sent to the hospitals in Edmonton and Fort McMurray.[285]

In the midst of all this, Suncor inadvertently dropped its own bombshell. While applying for approval to proceed with the construction of an open-pit mine at the Steepbank River site and its new Voyageur upgrader plant, Suncor filed a set of reports on chemical contamination affecting water, soil and wildlife in the area. Among the "chemicals of potential concern" listed in these reports were benzene, mercury and cadmium (also a carcinogen). Although arsenic was not tested for in these studies, the Golder report, which had been prepared by a consultant firm for Suncor in 2005, was tabled later in the hearings, predicting that the arsenic in moose meat — a staple of the diet for most area First Nations — could be 453 times the acceptable levels in the future. The Golder report came as a shock to the people of Fort Chip. As well, adds Chipewyan resident Archie Waquan, "If it's in the moose, it'll be in everything else we eat — rats [muskrats], cranberries and all that."[286] Not only were they unable to eat fish on a regular basis because of high levels of mercury and other toxins, but now it seemed that all of their basic land- and water-based food staples might be poisonous as well.

Thus, the tar sands boom is riddled with brewing social tensions and potentially explosive social conflicts. But this social upheaval is not confined to Fort McMurray, the epicentre of the tar sands, or even to within the borders of Alberta. On the contrary, as the development of the tar sands becomes more central to the Canadian economy, the more its problems will affect and spread to the rest of the country. The skewed labour pool, for example, that now exists across the country as a result of the tar sands companies' taking skilled workers from Newfoundland to British Columbia is likely to provoke more social tension, especially in times of a downturn in the Canadian economy. Similarly, conflicts may well start bubbling up in the manufacturing industries of Ontario and Quebec, where manufacturers are losing export markets for their products and workers are losing their jobs, largely because of the inflated value of the Canadian dollar, propped up by high oil prices and the tar sands boom itself. We return to this point in the next chapter. Meanwhile, there are a few other factors in the dynamic of the social upheaval that need to be discussed.

Energy Insecurity
The social upheaval generated by the tar sands is not a local phenomenon. A new wave of anxieties over the cost and use of Canada's oil and natural gas supplies to meet our own domestic needs, as well as over the prospect of increasing our use of nuclear power and other options, is beginning to emerge. And the juggernaut at the centre of these energy insecurities is the Alberta tar sands.

When it comes to extracting and producing oil from the tar sands, a great deal of natural gas is used for both the open-pit mining and in-situ drilling. According to the Pembina Institute, the open-pit mining operations require 21 cubic metres of natural gas to produce each barrel of oil. "In situ methods, which account for nearly half of the tar sands operations now and up to 80 percent in the future, require twice as much. It takes 42 cubic metres of natural gas to convert a barrel of bitumen into a barrel of synthetic crude that is light enough to transport to refineries."[287] By comparison, the average Canadian home heated with natural gas uses 255 cubic metres of gas per month during the winter. Therefore, current tar sands operations consume at least as much natural gas every day as is needed to heat half the homes in Canada for a day.

As several observers have noted: "It's like alchemy, turning gold into lead." After all, natural gas is a remarkably efficient fuel, and in many respects, a relatively clean fuel in comparison to other fossil fuels, despite the fact that it generates high volumes of greenhouse gas emissions. Unlike thick crude, it is easily transported through pipelines. For these and related reasons, many people in Canada converted their heating systems to natural gas following the oil crisis of the early 1970s. Now, according to Environment Canada, over half the homes in the country depend on natural gas for their heating. What's more, natural gas is the key raw material used in the production of a wide variety of other products, ranging from petrochemicals and pharmaceuticals to plastics and fertilizers. Small wonder, then, that this highly intense use of a relatively clean fuel with multiple uses to produce a dirty fuel primarily for export to the US is seen as "turning gold into lead."

To be sure, the production of all fuels uses energy. But, the question arises, is it worth expending a "clean" form of energy to produce a "dirty" one? The EROI (energy-return-on-investment) formula can provide one kind of answer. By applying this formula, one is able to assess the net energy gain or loss in fuel production. As energy analyst Richard Heinberg puts it: "The net-energy figures for tar sands are discouraging ... it takes the equivalent of two out of each three barrels of oil recovered to pay for all the energy and other costs involved in getting the oil from the oil sands."[288] Given the importance of natural gas for the heating of our homes during Canada's cold winters, coupled with the multiple uses of this resource in other products, Canadians have good reasons to question whether it is worth expending so much of our natural gas reserves on producing a synthetic crude when

the energy payoff is so small. Add to this the enormous environmental costs associated with tar sands production discussed in
Chapter 5 (boreal destruction, carbon emissions and water
depletion and contamination) and it is difficult to conclude that
there are significant energy gains to be made from the mining of
the tar sands.[289]

Worse still, from an energy security standpoint, Canada's conventional supplies of natural gas peaked in 2002 and are now in
decline. David Hughes, a veteran geologist who was with the
Geological Survey of Canada for more than thirty years and is
now at Natural Resources Canada, began studying Canada's natural gas supplies in 1996 and found patterns of shortfall. Hughes
says that Canada's natural gas supplies have remained relatively
stable since it peaked in 2002, primarily because of a massive
increase in the amount of drilling.[290] However, now there is very
clear evidence that the feverish pace of drilling is slowing down
and we can expect to see sharp declines in Canadian supplies of
natural gas very soon. Hughes has become increasingly concerned about the impact these declines will have on Canada's
long-term energy security.

According to Hughes's research, around 4,000 natural gas wells
were in production in 1996 in this country. By 2003, there were
more than 14,000 wells in production, but their overall productivity had declined by 3 percent.[291] A major reason, says Hughes,
is a decline in the productivity per well drilled. In 1996, the average gas well produced around 17 million cubic metres per day.
Today, the average output per well drilled is around 6 million
cubic metres per day. As a result, it is now necessary to drill two
to three wells "to get the initial productivity of a well that was
drilled back in 1996."[292] If Canada has colder winters over the
next few years, then the price of gas will increase, leading to an
acceleration in the pace of drilling. But more drilling does not
mean finding enough new gas to replace the gas being used. Since

the late 1990s, says Hughes, "our proven reserve base has been falling consistently," that is, we are running out of conventional supplies of natural gas.

Meanwhile, Canada's consumption of natural gas has been rising and exports to the US have been increasing. In 2005, Canada produced over 180 billion cubic metres of natural gas. Of this amount, more than 100 billion cubic metres, or roughly 60 percent, was shipped south of the border.[293] As America's domestic gas reserves continue to decline, the National Energy Board predicts that US demand for natural gas will reach 849 billion cubic metres per year by 2014. Since the Western Sedimentary Basin (which primarily covers Alberta and Saskatchewan) is the largest supplier of natural gas to the US, Canada is expected to export up to 164 billion cubic metres a year. On the domestic front, demand for natural gas is also growing. And the Alberta government predicts that natural gas consumption for tar sands operations will triple by 2015. In effect, one-third of all of the natural gas consumed in Canada will go towards tar sands production.[294] But then what will happen to the millions of Canadians who depend on natural gas to heat their homes?

As of 2005, the Canadian Centre for Energy Information reported that, given current rates of extraction, Canada's conventional natural gas reserves will supply fewer than ten years' demand. As of 2008, this means that there are just something less than eight years of supply remaining in conventional natural gas reserves, a fact that serves to heighten tensions around the issue of energy security in this country. Without sufficient supplies of natural gas to meet domestic needs, not only will people be unable to heat their homes during the winter, but thousands of industries across the country who depend on gas to produce their goods and services may be compelled to leave altogether.

For some, the answer to this dilemma lies in the High Arctic islands, now rapidly being denuded of ice by global warming. In

2008, Natural Resources Canada issued 1.2 billion dollars' worth of exploration leases to transnational oil and gas companies for exploration on Canada's High Arctic islands, until recently considered to be too rugged and costly a terrain to develop. But because of the twin evils of global warming and skyrocketing energy prices, they are destined to become the new tar sands.

For others, the answer is the Mackenzie Gas Project (MGP) and the proven Beaufort Sea gas reserve. The goal of the MGP is to build a 1,400 kilometre pipeline to transport between 33 and 54 million cubic metres of natural gas per day from the Mackenzie Delta on the Beaufort Sea through the Mackenzie Valley in the Northwest Territories into northern Alberta. Estimates of how much natural gas there is in this region vary from between 680 billion to 1 trillion cubic metres to as high as 1.8 trillion cubic metres.[295]

In any case, the proposed Mackenzie Gas Pipeline corridor is primarily for the delivery of gas to the Alberta tar sands. Former premier Ralph Klein confirmed this in a speech he delivered at Harvard University in the spring of 2005, saying the MGP would provide the tar sands with a twenty-year supply of natural gas.[296] The MGP has been planned by a consortium of corporations led by Imperial Oil Resources Ventures Ltd., a subsidiary of Imperial Oil Ltd., whose parent corporation is the US oil conglomerate, ExxonMobil. Imperial Oil was originally to construct and operate the gathering system for natural gas in the Mackenzie Delta where it has its Taglu gas field. The consortium includes the Aboriginal Pipeline Group Ltd. (APG), composed of representatives from the Inuvialuit people of the Mackenzie Delta and Arctic coast, the Sahtu First Nation from around Great Bear Lake, and the Gwich'in First Nation who live south of the Inuvialuit. Backed by a loan from TransCanada Pipelines Ltd., the APG was organized to represent the interests of some of the Aboriginal peoples of the NWT. The other members of the consortium are US and international oil giants through their subsidiaries, namely,

ConocoPhillips Canada, which operates the Parsons Lake gas field, and Shell Canada, which runs the Niglintgak gas field.

However, it remains uncertain as to whether the MGP will be built in time to service the increasing production schedule of the tar sands and offset the growing demand for natural gas in the country, let alone exports to the US. With delays in both the National Energy Board hearings and the Joint Review Panel reports, the start-up costs have ballooned. The price tag for the project has shot up from 7.5 billion to almost 19 billion dollars, threatening the feasibility of the enterprise. At its annual general meeting in May 2007, ExxonMobil's chief executive officer warned that the Mackenzie Gas Project was not economical given current costs and would not be built without significant federal aid. Alternative financing proposals were floated, including a plan for TransCanada Corp. to assume 60 percent ownership, and others involving making the federal government a joint partner and owner of the project. At one point, ExxonMobil threatened to pull out altogether unless Ottawa would help to defray mounting costs. But industry minister Jim Prentice, who has been a strong advocate of the gas pipeline, declared in December 2007 that the Harper government had no interest in owning any portion of the project or "in subsidizing the petroleum companies."[297] Meanwhile, ExxonMobil has given greater priority to the building of the Alaska natural gas pipeline from Prudhoe Bay, thereby redirecting much of the labour and steel needed for the MGP.

At the same time, other options are being explored. One is coal-bed methane, a natural gas found in most coal deposits. As author and journalist Hugh McCullum explains: "The methane in a coal seam is not stored as a compressed gas, but is absorbed chemically into the coal and held in place by the overlying rock and water pressure."[298]Although Canada has the twelfth largest coal reserve in the world, there is no way of accurately determining how much gas there is, and how much of that is recoverable.

And the process used for extracting the methane gas from the coal bed, like those used to extract bitumen from the tar sands, requires a great deal of water. In the United States, where they have had a twenty-year history of developing coal-bed methane, serious problems with the depletion and contamination of aquifers and with wastewater disposal have emerged, making "the practice of removing methane from ancient coal seams a controversial and emotion-laden issue." Although in 2005 the National Energy Board predicted that the industry would be able to pump between 13 and 23 percent of known deposits of coal-bed methane by 2025, geologist David Hughes of Natural Resources Canada says that this will not compensate for the declines in conventional natural gas.

There is another option for fuelling the tar sands that's gathering more momentum: nuclear power. In October 2006, Atomic Energy of Canada Ltd. announced it had signed an exclusive contract with a start-up Calgary-based company, Energy Alberta Corporation, to utilize the CANDU reactor technology in the tar sands. Canada's Natural Resource Minister, Gary Lunn, is a strong advocate of this option: "I think nuclear can play a very significant role in the oil sands."[299] It also creates a potential standoff between two ministries: the nuclear energy option promoted by Lunn versus the natural gas option promoted by industry minister Jim Prentice.

In August 2007, Energy Alberta announced that Peace River had been chosen as the site for its new nuclear power plant. A pair of twin-unit CANDU reactors are to be built on property adjacent to Lac Cardinal. While nuclear energy is cleaner than natural gas in terms of greenhouse gas emissions, there are a host of other issues raised by the reactor itself, as well as issues around the storage of waste and transportation of nuclear materials. In addition, says University of Alberta's David Schindler, nuclear power plants use a lot of water for cooling. "This is one reason for putting it [the new Alberta nuclear plant] in the Peace River, so they can get the water from the Peace. The needs are around a cubic metre a second, so it's like a small oil sands plant."[300] And the Concerned Citizens Advocating Use of Sustainable Energy warns that the proposed Alberta plant is "a different beast" from nuclear plants in Ontario, Quebec and New Brunswick, which are half the size of the twin-reactor plant being proposed.

With the tar sands virtually next door, neighbouring Saskatchewan is also vying for a piece of the action. As a major supplier of uranium used by nuclear power plants around the world, Saskatchewan is in a perfect position to develop nuclear energy options in the tar sands. A significant portion of the world's known uranium resources are located in Saskatchewan. The uranium deposits in Saskatchewan are large, contain high-grade ore and can be extracted at production costs below those in many other parts of the world. Saskatchewan's uranium resources are sufficient for more than forty years at current rates of production. Since the 1950s, sixteen ore bodies and three separate milling facilities have been developed in the Uranium City area. More recently, the TransCanada Corporation, which owns a large share of Bruce Power, one of the largest producers of nuclear energy in Ontario, has been actively promoting the development of nuclear plants in Saskatchewan as well as Alberta. Nuclear in Saskatchewan makes sense, says TransCanada's CEO Hal Kvisle.[301]

Following his election as Saskatchewan premier in 2007, Brad Wall has been positioning his government to move quickly on the building of at least two nuclear power plants: one to supply electricity to other provinces and the other to provide power and steam for tar sands developments.

In April 2007, however, a House of Commons parliamentary committee issued a report saying that plans to use nuclear power plants to supply Alberta tar sands production should be put on hold until the full consequences are known. According to the Commons committee, it would take the construction of almost twenty nuclear reactors to fuel the tar sands plants for the increased production planned between 2007 and 2015.

Other parts of the country are becoming increasingly vulnerable with respect to secure energy sources. While the Western Sedimentary Basin provides parts of Ontario as well as the Prairies and British Columbia with their oil and natural gas needs, Quebec, the Atlantic provinces and a third of Ontario are highly dependent on imported oil. The east-west pipeline system that used to connect the Alberta oil fields with the rest of the country no longer goes beyond Sarnia. As a result, the eastern provinces, comprising over one-third of the entire country, get 90 percent of their oil, 850,000 barrels per day, through imports. The Sarnia to Montreal pipeline that was opened under the Trudeau government to deliver western oil to eastern Canada was reversed in 1999. The pipeline now flows from Montreal to Sarnia, delivering some of the oil imported into eastern Canada through southern Ontario to the Michigan border. However, Enbridge Pipelines Ltd. announced in early September 2008 it plans to reverse the pipeline flow once again. Tar sands crude would be transported from Alberta through Sarnia to Montreal, not to provide energy security for eastern Canada, but to deliver crude to Portland, Maine, where it will be shipped by tankers to Gulf Coast refineries. Called "Trailbreaker," this project is expected to be completed by 2010.

In some ways, the Alberta bumper sticker that was popular during the turmoil around the NEP, "let the eastern bastards freeze in the dark," takes on new significance in the light of these facts. After all, half of the homes in eastern Canada are heated with oil and many industries in this part of the country cannot operate without it. Moreover, most of the imported oil comes from three OPEC countries — Algeria, Iraq and Saudi Arabia —

none of which can be counted on to be secure suppliers.[302] At any time, the oil supply chains from any of these countries could be cut off as a result of natural disasters, terrorist attacks or economic blockades. Indeed, we are living in an era when short-term oil supply shocks are becoming more and more common. If any of these oil supply lines were to be disrupted, the energy security of Quebec and the Atlantic provinces would surely be in jeopardy. The crisis would be all the more severe if this were to occur when eastern Canada was in a winter deep freeze.

What's more, Canada does not have a strategic petroleum reserve to offset oil-shock waves as other countries do. As the Parkland Institute underscored in its joint report with the Polaris Institute in January 2008, *Freezing in the Dark*, most industrialized countries in the world today keep a ninety-day supply of oil in reserve for emergencies such as oil supply shocks. After the shock of running out of oil during the First World War, France took the lead in 1925 by establishing what became the first strategic petroleum reserve. By 1968, most other western European countries had followed suit. In 1977, the United States established its own strategic petroleum reserve of more than a ninety-day supply stored in salt caverns along the Gulf Coast near key refineries. Even some net oil-exporting countries, such as Iran, Mexico, Russia and Saudi Arabia, have strategic petroleum reserves in case of emergencies. More recently, China and India have also begun to establish reserves.[303]

But not Canada. The International Energy Agency, which oversees the energy policies of all non-OPEC countries, requires all its members to maintain an emergency oil reserve. It exempts only those countries that are net exporters of oil on the assumption that they will meet their domestic needs before transporting their oil to other countries. But Canada does not meet the IEA criteria for the exemption. While we are a net exporting country, we no longer have the Canada-first policy we once did whereby a twenty-five-year supply was in reserve before exports were allowed. Today, 67 percent of our oil and 59 percent of our natural gas is exported to the United States, but under the NAFTA proportionality clause (as we saw in Chapter 4), Canada is forbidden to cut back or put a quota on our oil exports to the US even if we encounter serious shortfalls in domestic supply. And even if Canada were able to renegotiate this part of NAFTA in order to be free to implement a Canada-first policy to ensure energy security for all its regions, the infrastructure required to store a reserve is no longer in place.

Canada's vulnerability in terms of energy security was underlined by the response University of Alberta professor and president of the Parkland Institute Gordon Laxer received in April 2007 from the National Energy Board (NEB) regarding a series of inquiries he had submitted in preparation for his report *Freezing in the Dark*: "Unfortunately, the NEB has not undertaken any studies on security of supply."[304] Yet, says Laxer, the NEB was established back in 1959 with precisely the mandate of ensuring the long-term security of Canada's energy supply. A few weeks later Laxer appeared as a witness before the House of Commons committee on international trade to present the results of his research on energy security issues in Canada. But almost as soon as he began, his testimony was shut down by the chair of the committee, a member of the Harper government, who promptly adjourned the proceedings.

In short, there are serious grounds for concern about increasing energy insecurity in Canada, particularly in relation to supplies of oil and natural gas. If this country's conventional supplies of natural gas decline over the next few years, as projections indicate, heading for the danger point, we will not be able to establish a Canada-first policy and cut back on our exports to the US because of the proportional sharing clause in NAFTA. Nor do we have the pipeline infrastructure in place to immediately and fully respond to an emergency in eastern Canada should there be a sudden disruption to to the oil supply chains on which it, and two-thirds of our population, depend. The fact remains that we no longer have a made-in-Canada energy policy; rather, our energy "policy" is largely determined in Houston and by a market dominated by the big petroleum companies. And now, increasingly, the Alberta tar sands can be seen to be the linchpin keeping the wheels on the juggernaut of energy insecurity.

Our energy insecurity in Canada is bound to intensify in the current geopolitical climate of peak oil and global warming. The only way out of this potential national crisis looming on the horizon is to develop and implement a made-in-Canada energy policy, including a completely new strategy for tar sands development. This possibility is explored in more depth in Chapter 8.

US Military Links

Today, the social upheaval brewing over energy insecurity in Canada is accompanied by a growing concern over the use that tar sands crude is put to in the United States. The fact that the Pentagon is the single biggest consumer of oil in the US raises the concern that Canada has become America's military fuel pump.

The 2003 invasion of Iraq by the US-led coalition forces sparked an ongoing public debate about the connections between oil and war in today's global economy. After all, Iraq is a strategic source of high-quality oil that is relatively inexpensive to produce. When asked to participate in the invasion and occupation of Iraq, the Chrétien government said "no" to the Bush administration. Two years earlier, Canada had already sent troops into battle in Afghanistan, and in late February 2008, the Harper government, backed by the Liberal opposition, approved the extension of the mission until the end of 2011. Nevertheless, Canadians tend to be somewhat smug because we didn't go to war for oil. We like to chide the Americans, especially the US media, for denying the connection between oil interests and the war in Iraq. Yet, as the number one exporter of oil to the US, Canadians might want to ask whether we are complicit.

For many years, the Bush administration in Washington steadfastly denied there was any connection between the the decision to invade Iraq and its oil resources. In 2003, when Secretary of Defense Donald Rumsfeld was confronted with the question of whether the United States invaded Iraq to gain control of its oil, then one of the largest reserves of light crude in the world, he flatly replied that was "utter nonsense. ... We don't take our forces and go around the world and try to take other people's ... resources, their oil. That's just not what the United States does." Two years later, speaking to US troops in Fallugah, Iraq, Rumsfeld reinforced his point about the purpose of the military mission: "The United States, as you all know better than any, did not come to Iraq for oil." In late 2006, however, around the time of the midterm elections, when the American people were expressing their growing resistance to the war in Iraq, President Bush himself more or less indicated "the war is about oil" when he said: "You can imagine a world in which these extremists and radicals got control of energy resources."[305]

There are important connections between the invasion of Iraq and the tar sands boom in Alberta that are worth drawing out. Before the invasion of Iraq in April 2003, the worldwide price of oil was high enough to make the production of tar sands crude economically feasible, but not terribly profitable. It currently costs less than 25 dollars a barrel to produce crude oil from existing tar sands plants. In 1998–99, the world price per barrel of oil hovered around 12 dollars. By March of 2003, the price of oil had jumped to 35 dollars a barrel. As well, 2003 was the year that the United States Energy Information Administration officially recognized that the Alberta tar sands contained one of the largest hydrocarbon deposits of "economically recoverable" oil on the planet. Following the invasion of Iraq, the Middle East was further destabilized and oil supplies tightened up around the world, thereby creating conditions for oil prices to skyrocket.

As author and journalist Naomi Klein wrote in October 2007: "For four years now, Alberta and Iraq have been connected to each other through a kind of invisible seesaw: As Baghdad burns, destabilizing the entire region and sending oil prices soaring, Calgary booms."[306] The initial US plan, originally conceived before the events of 9/11 by the Project for a New American Century, called for the invasion of Iraq and the toppling of the Saddam Hussein regime in order to secure control over Iraq's less expensive, high grade oil for the major US petroleum corporations. Not everything, however, went according to plan.[307] The prolonged resistance of Iraqi insurgents, who often targeted their country's oil fields and infrastructure, coupled with ongoing delays in the passage of Iraq's new oil law, made the US oil giants reluctant to carry out their role in the plan. Without Iraqi oil coming online, access to cheap oil was disrupted. Worldwide demand continued to increase and available oil supplies shrank, pushing prices into an upward spiral well past the 100 dollar per barrel mark.

With a strange twist of irony, the burning of Iraq's oil infrastructure has, in effect, triggered the booming of the Alberta tar sands. However, the military connections and our complicity run even deeper. By currently exporting 1.2 million barrels a day of crude oil to the US from the tar sands alone, with plans for a fivefold increase by 2020, Canada is rapidly becoming America's military fuel pump. After all, the Pentagon is the single largest institutional consumer of oil in the US, and, for that matter, the world. Everyday, the Pentagon reportedly consumes and burns 365,000 barrels (some estimates put the figure closer to 500,000 barrels a day). This is 85 percent of the US government's total oil consumption and is equivalent to the daily oil consumption of the entire country of Sweden. Prior to the invasion of Iraq, the

Pentagon's official figure for annual oil consumption was 110 million barrels. Since then, the Pentagon admits to burning 130 million barrels a year.[308]

Simply put, the Pentagon needs oil in order to survive. Altogether, the jet fighters, bombers, tanks and other vehicles used by the US military burn 75 percent of the fuel used by the Department of Defense. B-52 bombers, for example, require 1,119 barrels for each of their military missions over Afghanistan. Every time an F-16 fighter jet takes off, it burns through roughly 300 US dollars' worth of fuel per minute.[309] In Iraq alone, the US Department of Defense has 27,000 vehicles, most of which get lousy gas mileage. The M-1 Abrams Tank, for example, gets less than less than half a kilometre per litre, while the Bradley fighting vehicle travels less than one kilometre per litre. According to the Defense Energy Support Center of the Pentagon, US forces consumed roughly 4.2 billion litres of jet fuel for their operations in Iraq over less than eighteen months (March 19, 2003 to August 9, 2004). On a monthly basis in 2005, the US military's aircraft, ships and ground crews were reportedly guzzling between ten and eleven million barrels of oil, primarily in Afghanistan and Iraq.

In retrospect, the Alberta tar sands boom turned out to be America's bonus prize for the Iraq invasion. Given the rapidly increasing fuel demands of the Pentagon, the expansion of tar sands production and crude oil exports have proven, as we have seen, to be essential for US national security interests. As one energy consultant put it, the Alberta tar sands have become "America's energy security blanket." Moreover, some of the major oil corporations operating in the tar sands also have highly lucrative contracts with the Pentagon to supply oil. For example, ExxonMobil, Shell Oil, and BP were listed among the Department of Defense's top contractors in 2006, receiving over 3.5 billion dollars from the Pentagon.[310] The first two have investments in tar sands plants that are currently operating, while the third has leases in Cold Lake and owns and runs one of the main refineries in the US (in Whiting, Indiana) for crude oil from the tar sands. Another major player in the tar sands, ConocoPhillips, had contracts with the Pentagon totalling more than 180 million dollars in 2006.

It is, of course, difficult to ascertain exactly how much tar sands crude, or for that matter how much of Canada's overall oil exports, are consumed by the Pentagon for military operations. Most of the information required to make such an assessment is classified as "top secret" for national security reasons. As described in Chapter 3, the flow of tar sands crude can be tracked as it moves through the maze of pipeline routes to refineries in the US to be refined and processed for a number of different purposes, and where it is also likely to be mixed with oil from other sources. Tar sands crude may be refined for use as jet fuel for airlines or gasoline for cars and trucks. Or it may be used to make various oil-based products. Tar sands crude may also find its way into uses in other industries, including military operations, or for heating homes.

According to the NEB, the lion's share of tar sands crude oil goes to the Midwestern states.[311] In 2005, 70.25 percent of tar sands crude went to this region through the Enbridge pipeline, which runs from Hardisty, Alberta, to Chicago, Illinois, and is used most heavily by Illinois, Indiana, Michigan and Ohio. The Rocky Mountain Region was the second biggest recipient of tar sands crude, 18.52 percent, most of it earmarked for Colorado and to a lesser extent Montana, Utah and Wyoming. The third main recipients of crude oil from the tar sands were the western states, notably Washington, at 6.2 percent. It is here where crude oil is refined for US military operations on the west coast. The Cherry Point Refinery, which is run by BP, is the largest west-coast supplier of jet and diesel fuel to the US military. Recently, BP announced it is making changes at its Cherry Point plant so it will be able to increase the volume of tar sands crude oil it refines.

The use of tar sands crude for US Department of Defense fuel needs was specified in the 2005 Energy Policy Act. As discussed in Chapter 4, a section of the 2005 Energy Policy Act, entitled "Use of Fuel to Meet Department of Defense Needs," specifically provides the US secretary of defense with the authority to develop a strategy that makes use of fuels produced from coal, oil shale and tar sands sources. However, the 2005 Energy Policy Act has since been superseded by the US Energy Independence and Security Act, signed into law by President Bush on December 19, 2007. This new law contains provisions that appear to contradict the clauses in the 2005 law regarding the use of tar sands crude for military purposes, and which may pose serious obstacles for exports in the future.

The 2007 US energy law prohibits the federal government from procuring fuels with a higher carbon content than conventional fuels. Since the bitumen from which tar sands crude is made has a higher carbon content than conventional crude, chances are it is not in compliance with this legislation. Michael Wilson, Canada's ambassador to the United States, wrote to Secretary of Defense Robert Gates, Secretary of State Condoleezza Rice, and Secretary of Energy Samuel Bodman in February 2008 warning that a narrow interpretation of the new legislation would have "unintended consequences for both countries." In January 2008, the powerful chairman of the House oversight committee in the US Congress, Henry Waxman, wrote to Defense Secretary Gates requesting a list of all Pentagon projects affected by the law and an outline of the steps that will be taken to bring them into compliance.[312]

In order to circumvent this requirement of the new US energy law, White House sources have indicated that the use of carbon sequestration technologies or the offsetting of carbon emissions with carbon credits might be sufficient to bring tars sands crude oil from Canada into compliance. In March 2008, Canada brought forward regulations mandating that after 2012 new tars sands plants would have to have carbon capture and storage technologies in place. But neither the Democrat Waxman nor his Republican counterpart on the House oversight committee find this acceptable. "The promise that in the future," says Waxman, "there might be ways to avoid increases in greenhouse gases would not be sufficient to meet the requirements." Besides, he emphasized, "When the government spends taxpayer dollars to fuel government operations, it shouldn't choose fuels that make the problem of global warming worse."[313]

For Washington insiders, the "greening of the Pentagon" is not entirely new. In 2005, then Defense Secretary Rumsfeld called on the Department of Defense to develop plans for using alternative power sources and energy-saving technologies. While the Pentagon has dabbled in the use of wind power at its naval station at Guantanamo Bay in Cuba, and in the use of hybrid fuel-electric motors in some of its vehicles, it has not been able to shake its addiction to oil. Every year, the Pentagon signs a new set of multi-million dollar contracts with petroleum companies to provide ongoing supplies of oil for its military operations. In March 2007, the Pentagon announced oil supply contracts with ExxonMobil, Shell Oil, ConocoPhillips and eleven other petro-giants worth 4 billion dollars, and another set of fuel contracts in September 2007 with British Petroleum, Chevron and five other companies for 1.4 billion dollars. Moreover, the US army cancelled its plans to introduce "hybrid-diesel humvees" and dropped plans to retrofit the Abrams tank with a much more efficient diesel engine. As well, the air force did not go ahead with plans to replace its aging aircraft engines with more fuel-efficient ones.[314]

To allow continuing and growing crude oil imports from the Alberta tar sands under the new US energy law, the Bush administration may well have to resort to granting a national security exemption. In other words, the White House could declare that tar sands crude is strategic to US national security interests and therefore is exempt from the requirements of the legislation. The grounds for doing so are rooted in the evolution of the US doctrine of national security: defending and protecting the vital interests of America. In 1998, notes Michael Klare, the US National Security Council defined "vital interests" as those interests which are "of broad, overriding importance to the survival, safety, and vitality of our nation." For much of the past century, "oil" has been one of those "vital interests" to be defended and protected by the use of military force, if necessary, when secure access to key supplies is in jeopardy.[315] It is quite plausible, therefore, that the president could exempt crude oil from the tar sands from the fuel cycle carbon content conditions of the new energy legislation.

The grounds for invoking such a national security exemption are also rooted in geopolitical realities. As the US experience in Iraq has shown, there are serious limitations to how far the US military machine can spread itself in defending the US's oil-security interests abroad. Although the US will continue to protect its interests in strategic areas of the world, its role as globo-petro-cop has become increasingly unsustainable. The relatively recent "discovery" of one of the planet's largest hydrocarbon deposits right next door, containing 176 billion barrels of recoverable crude oil and up to 2.5 trillion barrels, depending on the development of new technologies, is of vital interest to US national security. While producing crude oil from the tar sands is certainly more expensive, it is a much more secure source of energy in a nearby, friendly country, and must be controlled and protected for reasons of US national security.

In December 2005, the headline of a *New York Times* article read: "China Emerging as Rival to U.S. for Oil in Canada." The opening paragraph went on to say: "China's thirst for oil has brought it to the doorstep of the United States." As the article points out, US claims to Alberta or Canada in general as its "doorstep" is supported by the fact that American investment controls between 40 and 50 percent of oil in Alberta already. Hence, the US claims proprietary interest in the tar sands. China's growing interests are viewed not only as competitive but also as a provocation. So when Syncrude sends a trial shipment of oil to PetroChina or when the Chinese assume a 40 percent interest in the proposed construction of a pipeline from Edmonton to Kitimat, BC, US national security interests are threatened.

US national security interests also stretch northward beyond the Alberta tar sands to the Arctic and the Northwest Passage. For centuries, the building of a navigable route between the Atlantic and Pacific as an artery for trade and resource development has eluded us. But now, as global warming continues to shrink the sea ice in the Arctic and the Northwest Passage becomes free of ice during the summer months, it may soon be navigable for commercial traffic. The geopolitical struggle for sovereignty over the Arctic and the Northwest Passage is well underway. Russia, for example, has been staking out its claim, announcing it will open new ports on the Arctic Ocean as petroleum hubs for the twenty-first century. Canada has been asserting its sovereignty over the continental ice shelf by building air strips on the sea ice, installing electronic sensing devices, and laying the infrastructure for two High Arctic bases. Recently, the US has reiterated its long-held position that the Northwest Passage is an international strait. The US has strategic vital interests in ensuring that its position is upheld.

It should come as no surprise, therefore, to learn that Canada is becoming an important target for US military interests. Canada's deposits of strategic resources, such as oil and water, coupled with the strategic importance of the Northwest Passage, are of vital interest to US national security.[316] The homeland security regime established in the US following the events of 9/11 now exercises effective control over the movement of people across our borders, while continental trade regimes such as NAFTA and the SPP secure the free flow of resources such as oil from Canada to the US. In 2002, the CD Howe Institute warned that the US could use its military to seal the Canadian border if it felt its national security was threatened. "Although terrorism poses a real threat," wrote Canadian military historian J.L. Granatstein, "it is not the most serious crisis. The danger lies in wearing blinkers about the United States when it is in a vengeful, anxious mood ... The United States is deadly serious about homeland defence. The Americans will act, alone if necessary."[317]

It remains to be seen, of course, what will happen when the Bush regime comes to an end. If Barack Obama is elected president in November 2008, and the Democrats continue to hold a majority in both Houses of the Congress, then there will be some changes in military priorities. In terms of Iraq, for example, it is expected that an Obama administration will fulfill its commitment to withdraw US troops over a sixteen-month period after assuming office. Undoubtedly there will be other shifts in military and foreign policy as well. But, despite these and related changes, the US will remain a largely military-based economy. The military-industrial complex that former US president Dwight Eisenhower warned about fifty years ago will still be very much intact. And as long as the US remains addicted to oil, the Alberta tar sands will continue to be an American national security interest.[318]

CHAPTER 7

Resistance Movement

The ecological nightmare and the social upheaval that have erupted around the development of the Alberta tar sands have become the sparkplugs of a new and growing civil resistance. Although this movement is in its embryonic stage, it has taken root and is on the verge of spreading across the country and the continent.

In the building of any social movement, it's important that its epicentre be located as close as possible to the front lines of the struggle. For the tar sands, this is Fort McMurray and its surrounding communities, many of which are First Nation communities. But, so far, the primary players engaged in resistance against the tar sands are environmental and public interest organizations based in Alberta, mainly in Edmonton and Calgary. While these groups may not be located on the front lines in the Athabasca, Peace or Cold Lake regions, the building of a base of ongoing resistance in Alberta at large is significant in itself.

After all, the National Energy Program left a bitter memory of eastern domination that persists to this day. However, the struggle against the tar sands cannot be won in Alberta alone. As we have seen, the tar sands mega-machine reaches far beyond the borders of Alberta, deep into the US, as well as other parts of Canada. An effective resistance movement, therefore, not only requires a strong Alberta base, but a national and binational one as well.

Campaign Networks

In recent years, a number of Alberta-based groups have come together in response to the tar sands challenge, representing a diverse mix both in terms of organizational capacities and entry points. Some groups specialize in research on the impacts of the tar sands, while others focus more on public education on key issues and still others engage in lobbying of, and direct action against, the governments and corporations involved. Any social movement requires these and related skills. For each group, the entry point also varies. Some concentrate on environmental impacts such as water depletion, greenhouse gas emissions, and the destruction of the boreal forests; others take on economic issues such as jobs and wealth distribution while others work on the energy policy implications; still others focus on the social struggles of workers, farmers and indigenous peoples in the affected communities. Diverse starting points like these are essential for cultivating a social movement. However, a loose coalition of diverse groups is bound to be a mixed bag, containing contradictions and even pitfalls.

From the outset, the Alberta-based movement against the tar sands has been largely grounded in research carried out by the Pembina Institute and the Parkland Institute. Engaging mainly in environmental and social analysis, Pembina has published many key reports on the impacts of crude oil production from the

Alberta tar sands on issues ranging from greenhouse gas emissions to water to land permits and royalties. Their widely circulated pamphlet *Oil Sands Fever* initially helped to set the terms of the debate, and their website provides a variety of fact sheets and election tools on some of the key issues.[319] Based in Canmore and Calgary, Pembina staffers such as Dan Woynillowicz, Simon Dyer and Amy Taylor are frequently called upon as resource people for public education events and for public comment by the media. The Parkland Institute contributes research and analysis of tar sands issues from a more political-economic perspective. Based at the University of Alberta in Edmonton, the Parkland team includes Gordon Laxer, Diana Gibson and Ricardo Acuña, who also serve as resource people for the movement. They too have published major reports on the tar sands in regards to energy policy and wealth distribution and royalties. Parkland's work has been especially important in revealing the implications of the Alberta tar sands for Canada's energy strategy.[320] When either Pembina or Parkland releases one of its reports, the Alberta government, often the premier himself, usually responds.

In mobilizing public awareness and action, organizations such as Greenpeace and the Sierra Club have been on the forefront with their campaigns. Although both Sierra and Greenpeace have put a priority on mounting campaigns around tar sands issues in Alberta, they have different capacities and organizing styles. For its part, the Sierra Club has an overall campaign in Alberta calling for a "Time-Out" in the development of the tar sands. The prime focus of their campaign has been organizing in communities in areas associated with the tar sands production and its impacts. To date, Sierra's prime targets have been Fort Chipewyan, Marie Lake, Upgrader Alley, and Strathcona. In Upgrader Alley, for example, the Sierra team of Lindsay Telfer and Leila Darwish have been working with local farmers in building

This toxic "waterfall" was the site of the first civil disobedience action in the tar sands by Greeenpeace. Courtesy Greenpeace.

resistance against community water shortages caused by the proliferation of upgrader plants for producing tar sands crude.[321] At the same time, Greenpeace has been using its well-honed skills in direct action to mount resistance against the Alberta tar sands. Throughout the 2008 Alberta election, for example, Greenpeace activist Mike Hudema and his team targeted Premier Ed Stelmach, bird-dogging him every step of the way during his campaign. Actions such as a banner drop at a blue ribbon Tory party dinner drew front page headlines and focused public attention on the Alberta tar sands and its impacts.[322]

Given all the land grabs involving Aboriginal lands in the Athabasca, Peace and Cold lake regions to make way for the oil companies to develop the tar sands, one would have thought that indigenous peoples would have been leading the resistance. As noted in Chapter 1, both governments and the oil companies have played a role in swindling the First Nations of the region out of their Aboriginal land rights. And pockets of resistance are emerging. Working on behalf of the Indigenous Environment Network, Clayton Thomas Mueller reports growing unrest in some of the First Nation communities, such as Fort Chipewyan and Fort McKay, concerning the negative impacts of the expanding tar sands production on the water and wildlife.[323] Northeast of the Athabasca region, Chief Bernard Ominayak and the Lubicon First Nation are determined to oppose the construction of the Gateway Pipeline designed to cross their land in order to transport crude oil from the tar sands to Kitimat in British Columbia for shipment to the US or, possibly, China. And in the Mackenzie Valley of the Northwest Territories, the Dehcho First Nation has been protesting against what is happening in the tar sands because of the contamination and depletion of the Mackenzie River, which flows down from the Athabasca. Moreover, if construction begins on the Mackenzie Gas Project before the Dehcho have settled their land claim with Canada, their resistance is bound to intensify.[324]

The clear-cutting of the boreal forest to make way for the strip mines of the tar sands industry has been a long-standing cause for opposition. Working with conservationists, First Nations and industry, the Canadian Boreal Initiative (CBI) has played a pivotal role in influencing groups involved in forest protection issues to focus attention on the tar sands. In recent years, the CBI has produced at least six major research reports on the boreal forest.[325] It also works closely with the World Wildlife Fund (WWF) and the Canadian Parks and Wilderness Society (CPAWS). In addition to promoting protection of the boreal forests in the tar sands region, the WWF has become increasingly involved in issues of water contamination and water depletion as well. For its part, CPAWS has been engaged with governments and First

Nations in trying to establish large tracts of the boreal as national parklands wherein resource and industrial development would be forbidden, to be reserved in perpetuity. In this context, the Dehcho First Nation has reached a tentative agreement with Parks Canada, calling upon Ottawa to declare an expanded Nahanni National Park reserve covering the resource-rich South Nahanni River watershed in the Northwest Territories.[326]

Increasingly, organized labour is becoming an active part of the movement against the tar sands. While the prime issue for the Alberta Federation of Labour (AFL) is jobs and working conditions, the tar sands industry poses major problems for the unions by bringing in temporary foreign workers to undercut their members.[327] Although the tar sands industry is currently the country's number one hot jobs market, the AFL realizes that a boom-and-bust resource economy cannot provide secure employment in the long run. As a result, AFL president Gil McGowan has been pressing the Alberta government to develop an industrial strategy for the province that puts a priority on creating value-added industries such as petrochemicals, instead of simply producing crude oil for export to the US. In addition, the Communications, Energy and Paperworkers Union (CEP), which represents most of the energy workers in the tar sands industry, has called for a slowdown in tar sands production. In this age of climate change, the CEP has become increasingly conscious of the dirty image of the tar sands industry as a result of its greenhouse gas emissions and the damage it does to water supplies. Indeed, the CEP has a good environmental track record. It has been an active participant in the alliance of organizations that mobilized support for Canada's signing of the Kyoto Accord. And it was also active in pressing the National Energy Board to reject the Keystone Pipeline, designed to expand the delivery system for tar sands crude to the US in the coming years. If more labour organizations were to take a clear stand in opposition to the tar sands industry and work in collaboration with other opponents of the development, it would strengthen the movement considerably.[328]

Other Alberta-based groups are engaged in public education and action on tar sands issues. Public Interest Alberta (PIA), for example, is a progressive and broad-based alliance that includes educational, environmental, labour, child care, seniors and community organizations throughout the province. Through their environment committee, PIA organizes public education events around tar sands issues with its member groups, thereby reaching a broader audience.[329] The Prairies Region of the Council of Canadians has also been active on tar sands issues, working with its chapters across the province. Recently, the focus of the Council's educational outreach has been on the water impacts of the tar sands, energy policy concerns, and the implications of the Security and Prosperity Partnership. Other groups raise public awareness around tar sands issues, for example, Toxics Watch Society of Alberta joined with Pembina, Sierra Club and Prairie Acid Rain Coalition to launch a court case challenging Imperial Oil's Kearl Lake project for its failure to meet environmental standards. There have also been popular education tools, for example, the video *Toxic Alberta* by VBS.TV, which vividly portrays the ecological and social damage being wrought by the tar sands industry through a variety of lenses by including interviews with Albertans ranging from former premier Ralph Klein to homeless people on the streets of Calgary and Edmonton.

The difficulties of mounting public education and action campaigns in Alberta on issues around oil should not be underestimated.[330] There is a culture of resistance in Alberta to anything that resembles a critique of the oil industry or prevailing assumptions about wealth and power. In many ways, Alberta is the cultural epitome of an advanced industrial-capitalist society. Almost anyone you talk to has a relative or friend who works in the oil industry. It is the bread-and-butter business for most

families and individuals in Alberta. To openly challenge the oil
industry is offensive. Moreover, as tar sands campaigner Jessica
Kalman of the Polaris Institute, who hails from Calgary, put it:
"Albertans are inherently stubborn ... they don't like others
telling them what to do." In a sense, both Calgary and Edmonton,
in different ways, bear some of the traits of company towns where
social relations are largely dominated by the big company that
employs the town's residents, or, in this case the consortium of
corporations that dominate the Alberta oil patch. In this climate,
the terms "environment" or "labour" or "Aboriginal" are at times
viewed as "dirty" words. And the media, with some exceptions,
tends to reflect these values. The Sun newspaper chain, for example, is fond of labelling Pembina, Parkland and PIA as the "wacky
Ps" or the "loony Ps." Developing education and action strategies
to penetrate this cultural barrier is a major challenge for tar sands
activists in Alberta.

Nevertheless, opportunities for developing a counterculture
amongst tar sands activists have occurred in other forums. Take,
for example, the water forum, Keepers of the Water, held annually in Fort Chipewyan. Here the Mikisew Cree and the Athabasca
Chipewyan, along with elders from other indigenous communities, especially from the NWT and northern Saskatchewan, host a
public forum on water issues with scientists and activist
groups.[331] In 2007 scientists presented the results of their studies
on water contamination and health effects due to enhanced production in the tar sands, while Pembina used the event to mark
the completion of their "connecting the drops" tour of waterways
linked to the Athabasca River and the tar sands. The forum also
provided an occasion to sort out differences. For example, the oil
industry's participation in the Cumulative Environment Management Assessment (CEMA) was a stumbling block for some of
the First Nations. At the Fort Chip water forum in August 2007,
both the Mikisew Cree and the Athabascan Chipewyan

announced they would no longer be participants in CEMA.

But the resistance movement against the tar sands is by no means confined to Alberta. Although Alberta is the locus of front line campaigns against the tar sands, there are related resource development struggles taking place elsewhere. In neighbouring Saskatchewan, the new Saskatchewan Party government of Brad Wall, which is committed to pursuing the Alberta model of rapidly developing oil, gas, uranium and potash resources, is becoming the target of growing grassroots resistance in that province. The government of the Northwest Territories appears to be following the Alberta approach to resource development too in opening up its diamond, gold, lead, zinc, coal and natural gas deposits to transnational companies, thereby provoking resistance from the Dehcho and others. Meanwhile, as noted in previous chapters, workers in the manufacturing plants of Ontario and Quebec are beginning to see that the massive job loss in the manufacturing sector is tied to the boom in the tar sands and other resource commodities, which opens the way for resistance from this and other sectors of the economy.

Furthermore, the 2008 Alberta election demonstrated that the tar sands struggle cannot be won if the battle is restricted to Alberta. Despite the fact that the ruling Conservatives were at a record low in the polls at the beginning of the campaign and tar sands issues were well publicized throughout, Premier Ed Stelmach stayed the course and in the end pulled off a whopping majority. In short, the chance of curbing, let alone stopping, the tar sands development in Alberta through political means seems negligible. So Alberta-based groups have been building links and developing networks with activists in neighbouring provinces, across the rest of Canada and in the United States. As a result, action strategies are being developed, and significant pockets of resistance are being mobilized on a pan-Canadian and a pan-American basis, in order to curb, if not stop, the tar sands mega-machine.

Several of the Alberta-based groups engaged in tar sand struggles are part of pan-Canadian organizations, namely the Sierra
Club, WWF CPAWS, the CEP and the AFL. Along with Pembina
and the CBI, these groups also have a presence in Ottawa, providing the possibility of concerted action on Parliament Hill in
response to issues under federal jurisdiction. The Ottawa-based
Polaris Institute also does research on the tar sands companies
and pipeline routes, acts as a facilitator to bring others together
to make presentations to parliamentary committees, organizes
public education forums on tar sands issues and mobilizes electronic letter-writing campaigns through its website.[332] The
Canadian Labour Congress, which is also based in Ottawa, and
has affiliates across the country, is now in a position to take on a
more active role in responding to tar sands issues since a set of
resolutions on climate change was passed at the national conven-
tion in June 2008.[333]

Meanwhile Canadian churches, through their social justice
program KAIROS, are organizing visits to the tar sands and surrounding communities with international church leaders in
2009. They have also been spearheading a drive to terminate federal subsidies to the tar sands industry.[334] Environmental Defence
has been producing reports on the environmental and health
effects of the development of the tar sands, and developing an ad
campaign with other groups.[335] A group called Oil Sands Truth
has been working with community-based organizations and First
Nations on tar sands issues: its activities include holding an
annual conference of grassroots activists. Other groups in Canada are working in collaboration with their US counterparts to
mount a consumer action campaign against bank loans to the tar
sands industry, targeting one or more of the big five — Royal
Bank, Canadian Imperial Bank of Commerce, Bank of Montreal,
Scotia Bank and Toronto Dominion. In addition, some public
sector unions are exploring how they might use their pension

funds as leverage to put the squeeze on tar sands companies. And the Council of Canadians, Parkland and Polaris have been campaigning against the energy and water export and integration schemes in the SPP and are calling for the removal of the proportionality rules from NAFTA.

In the US, the Natural Resource Defense Council (NRDC) has been organizing market-based campaigns to get US companies and institutions to stop using "dirty" oil made from tar sands crude because of the high volume of carbon emissions produced in its processing.[336] One of NRDC's campaigns is directed at the US airline industry, in particular United Airlines, to stop the use of jet fuel made from tar sands crude at the Wood Refinery in Illinois. NRDC has also been monitoring those states that have adopted the California standards for carbon content in fuels to see if tar sands crude is being used by their industries, governments or institutions. According to California's new laws regulating the carbon content of fuels, imports of tar sands crude would be disqualified, not only because they may fail to pass the state's carbon content standards but also because so much greenhouse gas is emitted in its production (extracting, upgrading and transporting) that it fails the requirements set for the full fuel cycle. On another front, the NRDC has organized a campaign to prevent the US Defense Department from being exempted from the section of the 2007 Energy Independence and Security Act that prohibits the federal government and its agencies from purchasing dirty fuels, including tar sands crude, whose greenhouse gas emissions are three times higher than that of conventional fuels.[337]

Other US-based groups have targeted pipelines, refineries and investments. The Indigenous Environment Network has been tracking the planned construction of pipelines designed to transport tar sands crude from Alberta through US states, across Indian-held lands, and exploring what kinds of resistance might be organized.[338] In Indiana, a coalition of community groups is conducting "toxic tours" of a BP refinery that is to be retrofitted and expanded so that it can upgrade bitumen from the tar sands into synthetic crude. In their protest, community activists say the planned expansion would increase the refinery's greenhouse gas emissions to 5.8 million tonnes a year or the equivalent of putting 320,000 additional cars on US highways. In addition, the Rainforest Action Network (RAN) has conducted research to discover which US banks are making loans to tar sands companies and plans to mount a consumer campaign aimed at those banks, notably Citibank and JP Morgan Chase, in collaboration with groups working on Canadian bank campaigns.[339] In Washington, DC, Oil Change International has been working with NRDC, RAN, ForestEthics and other US groups in actions on Capitol Hill, including demonstrations against the Alberta government's tar sands promotion and marketing campaign.[340] A valuable addition to these activities could come from an alliance between labour and environmental groups in the US pushing for the renegotiation of NAFTA, and Canadian-based organizations advocating an overhaul of the energy chapter, including the proportionality rule.

Beyond these national and binational arenas, there is interest in action against tar sands oil in other countries as well. Gradually, the Alberta tar sands have been gaining worldwide attention, both as the world's largest hydrocarbon energy development and the single biggest capital investment project. Increasingly, climate change advocates in Europe and elsewhere realize that the major obstacle to Canada meeting its Kyoto commitments is the Alberta tar sands. It is also becoming evident that as long as production and export of tar sands crude continues, the US addiction to oil will be prolonged, possibly to the peril of the planet itself. Since two global oil giants, Shell Oil and BP, both based in Europe, are active in the tar sands, they are quite likely to become targets of international campaigns. Indeed, Friends of the Earth International has signalled interest in profiling Shell's involvement in the Alberta tar sands as part of its international campaign on the company. And while Canada does not export oil directly to Europe, it is conceivable that market-based campaigns could be organized to influence Canadian industries using tar sands crude to produce other products for export.

But building opposition to the tar sands in these three arenas is difficult because of the influence and money the oil industry has. In a provocative article, environment consultant, planner and northern activist Petr Cizek points out that some of the major environmental groups in the tar sands struggle are supported one way or another by the industry.[341] Cizek cites concrete examples in which groups actually changed their positions in response to oil companies that were providing them with support, or had their press releases checked by their oil-industry backers. Although Pembina has played a pivotal role in providing research and analysis for groups engaged in the tar sands campaigns, its own list of "partners and clients" includes some of the industry's major players, such as ConocoPhillips, EnCana, Husky Oil, Petro-Canada, Shell Canada and Suncor. According to Cizek, the CBI gets its money from the Pew Charitable Trusts, the foundation established by J. Howard Pew, the patriarch of Sun Oil, who pioneered the development of the tar sands and created Suncor. The Pew money, reports Cizek, is distributed by CBI through grants to other environmental organizations, notably CPAWS and WWF, and many other conservationist and Aboriginal groups.[342] The Pew money, he argues, has influenced the positions of these organizations on key issues.

While most activists would find this unacceptable, the fact remains that most citizen groups involved in tar sands struggles require considerable resources in order to do the work that needs to be done and do it effectively. For any social movement the politics of funding can present a conundrum. Here, the questions that need to be addressed are: How do campaign-oriented groups obtain the funds they need without sacrificing either their basic principles or their standards? Could funders pool their resources to be distributed fairly among the players in a resistance movement? How can agencies distributing funds be held accountable, not only to those who write the cheques, but also to the groups engaged in building the movement? Unless these kinds of questions are openly and honestly discussed, the chances of building and sustaining a dynamic social movement in Alberta and elsewhere against the tar sands are low.[343]

These are some of the main campaign networks in which resistance against the tar sands mega-machine is currently being organized, and some of the challenges they face. At present, they operate in three geographical arenas: inside Alberta; across Canada; and in the US. Taken together, they are the ingredients for building a dynamic resistance movement against the tar sands industry for the ecological and social harm it is causing in northern Alberta, and for the threat it poses to Canada's role in the twenty-first century. But more needs to be done.

Moratorium Platform

One of the strategic challenges this burgeoning resistance movement faces is the development of a common platform. While the plurality of positions, strategies and tactics must be recognized and respected, it is imperative that there be a common set of goals and principles around which the diverse groups can unite, in order to build a viable social movement for the long haul. Here, the question arises as to what is the unifying principle that provides the glue required to hold individuals and often disparate groups together. Opposing the tar sands on the basis of specific issues (e.g., greenhouse gas emissions, water contamination and depletion, energy security, mounting social costs, Aboriginal land grabs, etc.) is important for stirring people's passions and igniting activism, but it is not enough to develop and sustain a resistance movement. What is needed is a unifying theme or message that encompasses and transcends the variety of specific entry points. This unifying message needs to serve as a common platform to rally people's commitment to action.

The task of developing a common platform, however, is often easier said than done. Discerning the common threads that tie diverse issues together, let alone connecting different sectors, organizational styles and personalities closely enough to enable the pooling of resources and working together for a common cause, can be a difficult and daunting task. Finding a unifying theme with the power to transcend multiple diversities is crucial. To do so, it may be necessary to capture the historical moment in which we are living, especially the deeper forces driving the events. For example, the triple crisis of peak oil, climate change and resource depletion (e.g., water) are not only among the major drivers shaping the destiny of the planet in the twenty-first century, they are also (as discussed in Chapter 2) an integral part of the struggle over the future of the tar sands. While these themes will be taken up again later, it is instructive to look now at what has been done to develop a common platform for change in the debate over the tar sands.

There have been at least two attempts to pull groups together. In the fall of 2005, Pembina took the lead in developing a declaration called "Managing Oil Sands Development for the Long Term" which they co-signed with eleven other environmental groups (including WWF, CPAWS and the Sierra Club). Publicly released on the first day of December 2005, the declaration called for the creation of a network of protected areas and corridors near the tar sands area, the elimination of subsidies for the tar sands industry, and a binding integrated resource management plan for the region. It also urged that the tar sands be "carbon neutral [zero net greenhouse gas emissions] by 2020," by combining "on-site emission reductions and genuine emissions offsets." No mention, however, was made of a moratorium. But then former premier of Alberta Peter Lougheed, himself, used the term "moratorium" the following year in calling for a "slow down" or "pause" on new production in order to provide time to clean up the "mess" of problems that had erupted around the tar sands.

Two years later, in October 2007, groups active on tar sands issues gathered once again in Kananaskis, Alberta. By this time, several of the groups had called for a moratorium, including Pembina and the Sierra Club. While there was still no consensus on this occasion that the term "moratorium" should be used, there was agreement to a call for "no new approvals" of leases for production in the tar sands region.

A call for a moratorium on major oil and gas developments, however, is not unheard of in Canada. After conducting his Royal Commission Inquiry into the construction of the Mackenzie Valley Pipeline in the 1970s, Mr. Justice Tom Berger called on the federal government to impose a ten-year moratorium on the building of the natural gas pipeline through the Northwest Territories, and on oil drilling in the Yukon. In his report *Northern Frontier/Northern Homeland*, Berger outlined a series of solid reasons for the moratorium, ranging from the imperative need to settle the land claims of the Dene Nation before proceeding with the pipeline and related petroleum developments, to the threat to the fragile ecosystem of the north posed by major resource development projects. Berger came to his conclusions after conducting hearings in all of the northern communities that would be affected by the development, a series of technical hearings on the construction of the pipeline and resource development and public hearings in major cities across southern Canada. Since 1977, when the Trudeau government formally accepted the Berger Commission's call for a moratorium, no pipeline has been built from the Beaufort Sea to markets in southern Canada and the US.

Although there are significant differences between now and then to keep in mind, the idea of imposing a moratorium on major petroleum developments is certainly not without precedent.

Today, in the public debate over the Alberta tar sands, there have been three kinds of moratorium strategies proposed, each giving the term "moratorium" a different meaning, and each leading to different consequences. It is worthwhile taking a closer look at these three options to clarify the similarities and differences between them.

1. *Temporary Moratorium*: This is the position that was articulated by Peter Lougheed in the summer of 2006. It is based on the view that there needs to be a more orderly development of the tar sands. The government's procedures for granting leases and permits has gotten out of hand and the process is now in a "mess" and needs to be made more efficient and orderly. In addition, the rapid development of the tar sands is heating the economy of Alberta too fast and steps need to be taken to cool it down. To cool down Alberta's overheated economy as well as bring order out of chaos, the government needs to temporarily slow down production and introduce a "pause" in its granting of permits and leases for new projects in the tar sands. Thus, this call for a moratorium is focused on halting approvals in order to make time to bring order to the tar sands development process. Once order is restored, then the approval of new projects would resume. This option for a moratorium, therefore, is temporary and short-term.

The idea of a temporary moratorium along these lines seems to have gained favour with Albertans since it was initially promoted in 2006. Despite the results of the 2008 Alberta election, wherein the majority of Albertans voted for Ed Stelmach's Conservatives, who reject the option of a temporary moratorium, many are still open to this strategy. However, it does not address any of the burning issues raised by the Alberta groups engaged in the tar sands struggle. This moratorium strategy does not specify conditions for improvements in the manner in which the tars sands are being developed, or in the effects it is having on communities, the province, the country and the world. But what it could do is open the door for mainstream Albertans to consider the possibility of other options for the Alberta tar sands.

2. *Conditional Moratorium*: This is the position that perhaps comes closest to what the majority of Alberta groups mean when they use the term "moratorium." It also calls for no new approvals and no further expansion, but rather than being temporary, this version of a moratorium calls for a halt that may end up being for quite a long time. It is based on a set of conditions that would have to be satisfactorily met before new tar sands projects would be permitted. The list of conditions could include requirements regarding greenhouse gas emissions, water quality and preservation, boreal forest protection, energy security provisions, wealth redistribution measures, revised royalty regime, First Nations' lands compensation, development of value-added industries, no use of cheap labour, petroleum export quotas, and a requirement that tar sands crude not be exported for military use.[344]

To advance a conditional moratorium, however, it would be up to the groups that form the alliance to determine the range of issues to be covered and the specific demands to be suggested as conditions that must be met before development projects can resume. The stronger the set of conditions to be met, the greater the chances are that the moratorium would be long term. As well, the cumulative impact of multiple conditions would also ensure that the moratorium remain in place for a longer period of time. Yet, for a conditional moratorium like this to work effectively, there must be a high degree of solidarity amongst the partner groups and stakeholders in the alliance. If, for example, environmental groups get some of their basic demands met, then it's imperative that they are committed to the continuation of the moratorium until the demands of First Nations, labour unions, community groups and other public interest organizations are satisfactorily addressed and resolved. Once the links in the chain are broken, the conditional moratorium falls apart. What's more, the conditional moratorium is still focused on "no new approvals" and does not deal with the damage already being done by existing tar sands projects.

3. *Permanent Moratorium*: This is a position held by a small minority of Alberta groups. Of the three moratorium strategies, it is certainly the most radical. It calls for the phase-out and eventual shutdown of the tar sands industry. The permanent moratorium position is based on the premise that the tar sands mega-machine is well on its way to causing irreparable ecological and social harm. Not only is it unsustainable, but it perpetuates the ongoing addiction to oil in our society and impedes the progress that must be made towards a non-fossil fuel energy future. This moratorium strategy includes target dates for the phasing out of existing tar sands projects with the goal of complete shutdown of the industry sometime between 2015 and 2020. In the meantime, no new approvals would be granted for expansion of tar sands development. This position includes job transition strategies as well as plans for transition to alternative energy sources.

Although no single Alberta group has fleshed out a comprehensive plan for a permanent moratorium, it is the subject of discussion in several organizations. But the permanent moratorium strategy faces a number of hurdles. The fact that the tar sands are the engine of the Canadian economy makes it difficult, if not impossible, to sell the idea that the industry be shut down. As noted above, there is reason to believe that most Albertans would dig in their heels and defend the right of the tar sands industry to exist and flourish at all costs. It is plausible, however, that the tide of public opinion could move in favour of a permanent moratorium and shutdown of the tar sands industry, if the impending crises over climate change and water depletion and contamination take a turn for the worse. Even so, to win broader support, the advocates of this strategy would need to come up with a viable jobs transition plan for existing workers and their families plus a green jobs plan for a new generation of workers. To be effective, it would also have to include a viable plan for the transition from fossil fuels to renewable energy sources.

Today, all three of these moratorium proposals are still in play as far as the majority of Alberta groups are concerned. Although most of the Alberta groups discussed here are likely to favour some variation of the second option, the conditional moratorium, the role that could be played by the other two options should not be dismissed. The first option at least holds open the door to dialogue with a broader segment of Albertans about the need for a moratorium, and it affirms a "no new approvals" position, albeit on a temporary basis. The third option keeps the door open to a more radical position, which could well gather increasing support as the damage done by expanding tar sands production becomes more widely known or if certain ecological or geopolitical crises intensify dramatically. Instead of choosing one option to the exclusion of the other two, it might be wise to keep all three proposals in play. By allowing all three options to engage in dialectical interplay with one another, the chances of momentum building around the call for a stronger moratorium on tar sands production could increase.

At first glance, the task of crafting a common platform that encompasses all three options for a moratorium may seem difficult. But it can be done. In forging a common platform, groups engaged in various tar sands struggles could put priority focus on one of the three options, say, the conditional moratorium. At the same time, however, there could be explicit recognition of the other two moratorium strategies, along with an expressed willingness to continue dialogue with groups and individuals in the other two camps. In dialogue with the first camp, the objective would be to emphasize the limitations of a temporary moratorium, and encourage them to take on one or more of the multiple issues dealt with in the conditional moratorium, thereby broadening the base of support for it. In dialogue with the third camp, the objective would be to explore a deeper set of reasons for maintaining an ongoing permanent moratorium on tar sands production while, at the same time, addressing some of the difficulties of the conditional moratorium position. Although the conditional moratorium position may be the best hope for a basis of unity, it's important to keep the political interaction between the three positions alive. Meanwhile, two federal political parties, the New Democrats and the Greens, have variously called for a moratorium on new tar sands projects.

A common platform for social change (whether along these or other lines) cannot be developed and implemented without the participating groups having a common forum for sharing information and engaging in discussion and debate. The kind of gathering that took place in Kananaskis in October 2007, which included Albertan, pan-Canadian and pan-American groups, could be formalized with regularly scheduled meetings taking place two or three times a year. To ensure effective outreach, however, this forum would need to include all those groups engaged in struggles against the tar sands, not only the major research and environmental groups, but also the First Nations, labour unions, public interest organizations, faith-based networks, concerned farmers and community-based groups. When people from diverse sectors come together in a common forum they have the opportunity to challenge one another and ultimately learn from one another. It is through this kind of process that the development of, and commitment to, a common platform calling for a moratorium on the tar sands can be strengthened and enriched. A common platform, in other words, requires a dynamic common space in which activists can cultivate trust and confidence in discussing and debating issues, thereby increasing their capacity for a politics of collective action.

In short, building a resistance movement to tackle the tar sands over the long haul means uniting like-minded activist and community-based groups from diverse sectors in a common cause. To do so requires developing a common platform for social change around unifying themes and issues that can be used to inspire and mobilize increasingly broader segments of the public. In order to develop and sustain a common platform, however, activists need a common space in which to come together on a regular basis, to share information on their respective activities, discuss and debate strategies, and cultivate their capacity for collective action. This is essential to meeting the challenge of forging a counterculture in Alberta, which is imperative in order to take on the tar sands mega-machine. But building a resistance movement requires more.

Resistance Building

As with any complex set of issues, curbing the tar sands mega-machine requires a multi-pronged campaign strategy. In the coming years, campaigns will need to be mounted on a variety of fronts around a series of critical issues. The campaign strategies and issue priorities will no doubt differ, depending on the kind of organization conducting the campaign, and the arena in which it is conducted (i.e., Alberta, pan-Canadian, or pan-American). By mounting resistance on a series of fronts, the objective is to throw monkey wrenches into the production and export plans of the tar sands industry and government officials in both the US and Canada. In so doing, the momentum for curbing tar sands production will grow, thereby consolidating public support for a moratorium. The time frames for campaigns will also differ depending on the issue, the group and the arena. But looking ahead three to five years, campaigns are likely to be organized around the following issues in each of the three main arenas.

In the Alberta arena, the contamination and depletion of fresh water by the tar sands industry will increasingly become a hot-button issue for resistance. Community, environmental and First Nations groups are currently organizing action campaigns on water issues in both the Athabasca region and Upgrader Alley, also known as "Cancer Alley." Energy security issues in the province, such as the rapid depletion of natural gas reserves and the problems associated with alternatives such as coal-bed methane and nuclear power, could well flare up into focal points for resistance by community and environmental groups. We can expect that environmental groups will continue to pressure tar sands plants and companies about greenhouse gas emissions and carbon sequestration, and that this pressure will grow. At the same time, the failure of the tar sands companies to satisfactorily complete the requirements for the reclamation of landscapes ripped apart by their strip-mining operations could well surface as a hot-button issue. The use and abuse of foreign workers could also become a flashpoint for labour union resistance in the province, and a new phase of the debate over royalties could open up, especially if the social upheavals generated by the tar sands intensify.

In the pan-Canadian arena, the role of the tar sands industry as the country's number one global-warming machine means that it will increasingly become the prime target for climate change activists, particularly in provinces where carbon emissions have become a major issue, such as Manitoba, British Columbia, Ontario and Quebec. Labour unions representing workers in manufacturing industries, especially in Ontario and Quebec, who have been hard hit by job losses due to severe cuts in the exports of their products because of the inflation of the Canadian dollar spurred on by high oil prices and the tar sands boom, could be mobilized for resistance. Issues of energy security affecting eastern Canada, (i.e., a third of the country's population) could erupt in the next few years because the tar sands oil boom is geared for exports to the US rather than serving the energy needs of people living east of Ontario. Community-based resistance could also emerge around the construction of the Gateway Pipeline from Edmonton to Kitimat. As well, the calls on Ottawa for a moratorium on the tar sands are bound to intensify around water and other issues noted above.

In the pan-American arena, campaigns against the import of tar sands crude on the basis that it is a dirty fuel can be expected to escalate. Already, US-based campaigns have been effective in getting the US Mayors' Association and several US states to undertake measures to ensure that tar sands crude is not being consumed by their cities and states. Major industry consumers, such as the airlines, will also be targeted for escalating actions by environment groups. Another critical issue for US groups is breaking the pattern of US oil addiction by demonstrating that increasing dependency on tar sands crude from Canada simply prolongs the addiction and obstructs the transition to a non-fossil fuel energy future. US Aboriginal groups (supported by farmer and environmental groups) will also be building resistance against the construction of the Keystone and related pipelines designed to transport tar sands crude across Indian-held lands to US markets. At the same time, community-based resistance campaigns are being organized against the construction and/or expansion of refineries in the US that are being planned for the processing of tar sands crude.

Together, the campaign strategies mounted in the three arenas around these and related hot-button issues could ignite a groundswell of resistance to the tar sands and a collective call for a moratorium. To do so, the campaign activities undertaken in one arena need to be reinforced by campaigns in other arenas and vice versa. This calls for a degree of coordination of campaign activities between all three arenas and a communication strategy for common messaging to generate more widespread public support. In the meantime, the resistance movement has been developing and growing around at least ten fronts. The following is a brief outline of these ten fronts of education and action on tar sands issues, along with a few suggestions on how each might be further developed and strengthened.

1. Public Front: Various public forums, teach-ins and action workshops have been organized on tar sands issues in cities and towns throughout Alberta. In most cases, these education events are geared to move a broader spectrum of the public from awareness to action on tar sands and related issues. But outside of Alberta, little public awareness about the tar sands and its implications has been developed so far. In order to build a broader public base of awareness and action, special attention needs to be given to making links between the critical issues at stake in the tar sands and related issues and interests in other regions of Canada and the US. Coupled with the other action fronts below, these more education-oriented initiatives will be important for generating a broad-based public momentum against the tar sands mega-machine.

2. Community Front: A growing number of communities on the front lines of the tar sands machine, including First Nations communities located downstream from Fort McMurray and farmers in Upgrader Alley, are organizing their own forms of resistance. Community resistance is also building against the infrastructure being built for the expansion of the tar sands industry and its operations such as the Keystone and Gateway Pipelines and the refinery expansions in Whiting, Indiana and South Dakota. These actions involve Aboriginal peoples, farmers, workers and environmentalists. Building effective resistance in communities where upgraders, refineries and pipelines are being constructed obstructs the tar sands industry in delivering their product to markets.

3. Consumer Front: Institutionally focused consumer campaigns are being organized to ensure that low carbon fuel standards, initially pioneered in California, and now adopted in ten other states, are rigorously enforced. Most of these campaign initiatives are underway in the US where the bulk of tar sands oil is being marketed. The objective is to get major airlines to stop using tar sands crude for their jet fuel, transport companies to stop using this dirty fuel for their truck fleets or cities to ensure their transit systems make use of cleaner alternative fuels. Similar consumer-action strategies are also needed in Canada, especially in Ontario, British Columbia and Manitoba, which have adopted similar low carbon fuel standards. These kinds of consumer campaigns serve to foreclose the marketing options of the tar sands industry.

4. Legal Front: Some legal actions are underway to put pressure on the tar sands industry and the provincial or federal governments. Although the grounds for environmental litigation are not as strong in Canada as they are in the US, First Nations groups are in a potentially better position to initiate legal actions based on the Aboriginal rights and title provisions enshrined in Canadian law. In Fort Chipewyan, for example, First Nations groups are exploring taking legal action in response to the contamination and depletion of their water due to the operations of the tar sands industry. In addition to Aboriginal rights law, the Fisheries Act of Canada contains water protection provisions that could be used as a basis for legal action. While relying on the courts has its limitations, litigation is a tool that needs to be explored further as a means for building pressure on both governments and industry.

5. Legislative Front: Enforcing legislation that already exists and creating new legislation where possible is an ongoing action priority for reining in the tar sands industry. In Canada, environment groups are taking action to ensure that the absolute emissions targets within the recently adopted Climate Change Accountability Act, initiated by New Democratic leader Jack Layton, are rigorously applied to the tar sands industry. Alberta-based groups are keeping a close watch on the Stelmach government and doing what can be done through the opposition parties to push for legislative action to regulate water takings and land reclamation by the tar sands industry. Similar efforts are taking place in Ottawa through the parliamentary committees on the environment and on natural resources, as well calling for a made-in-Canada energy policy that includes a quota on tar sands crude exports.

6. Financial Front: New campaigns are being organized to put pressure on the banks and financial institutions that provide capital to the tar sands industry. In Canada, this includes five of the major banks (Royal Bank, CIBC, Toronto Dominion, Bank of Montreal, Scotia Bank) that have done business with tar sands companies. As part of this campaign, bank customers are provided with a formula and related tools they can use with their calculators to determine what percentage of their savings is going to finance the destruction in the tar sands. Similar campaigns are to be mounted in the US on financial institutions (e.g., Citibank, JP Morgan) providing capital loans for the tar sands industry and refinery expansions. The campaign objectives are to insist that the banks impose environmental screens on their loans, to show that the tar sands industry is not a safe place to be lending depositors' money, and to reduce the profitability of tar sands production by raising the cost of capital.

7. Labour Front: Another important arena for generating pressure on tar sands production is through organized labour. The availability and cost of labour is a factor of production that the tar sands industry cannot ignore. In Canada, the increasing interest and commitment on the part of the labour movement to at least curb the production of tar sands crude provides opportunities for maximizing labour action. Unions across the country are in a position to exert pressure against the tar sands industry's increasing use and mistreatment of cheap foreign labour, and against the gutting of Canada's manufacturing sector largely due to escalating oil prices and the inflated value of the loonie, and to pressure governments to develop just transition strategies for workers and their jobs in the event of a phase-out of production.

8. Media Front: So far, the resistance movement has been relatively successful in publicly tagging the tar sands with a negative brand through its media action. The stories about the health risks experienced by the people of Fort Chipewyan due to the contamination of their water were particularly compelling. The advantage taken of the deaths of 500 ducks in the Syncrude tailings pond illustrates the importance of being able to seize the media moment. Dramatic portrayals of both the human and ecological impacts of the tars sands through the media can certainly increase the pressure on the companies and affect their public image. While tours of the tar sands by celebrities and opinion-makers are planned to exert media pressure, more attention needs to be given to getting messages out through the media, especially on the economic, energy and social issues on people's minds that relate to the tar sands.

9. Investment Front: Organizing institutional shareholder actions to put pressure on the tar sands industry is also a component of resistance. Although in its embryonic stage, efforts are being made to bring institutions holding significant investments in particular tar sands companies together around common causes for action such as carbon emissions and water takings. Unions are beginning to look at their extensive pension funds' investments in tar sands companies and considering ways in which they could use this leverage to put pressure on the companies. Most of the activity on this front will take place with little or no public fanfare until there is a breakthrough. But if institutional shareholder actions are well organized and coordinated, with clear messaging via the business media, considerable impacts can be made on investor confidence in the tar sands industry.

10. Corporate Front: Several European organizations have expressed interest in mounting campaigns against two of the international players in the tar sands industry, namely, Shell and British Petroleum. The big oil corporations have been the target of several successful international campaigns waged by groups in Europe and North America with counterparts in Africa, Asia and Latin America. Although Canada does not export tar sands crude to Europe, products using oil derivatives could become the focal point for action. What these campaigns can do is cast an international spotlight on the tar sands in Canada and challenge the reputation of the industry. To be effective, however, it is best that international corporate campaigns be organized after a strong base for action has been built up both nationally and binationally.

If the groups actively engaged in struggles against the tar sands today on these and related fronts are to blossom into a full-fledged resistance movement in our times, then much more widespread public participation will be necessary. Although the Alberta tar sands have been featured in national newspapers and magazines, cross-Canada television and radio documentaries and occasionally on phone-in talk shows, they continue to be viewed as an issue of concern for Albertans. Outside of Alberta, the tar sands appear to be little more than a blip on the radar screens of most people in Canada, and less than that in the US. Even in this era of growing petro-rage at the gas pump, most people have not connected the dots between the escalating gas prices they encounter these days and the tar sands industry. In order for this resistance movement to grow, concerted efforts are needed to help people make the connection between the tar sands and issues that impinge directly on their daily lives.

In the meantime, here are a few basic steps that all of us could take in order to become active participants in this emerging movement.

- *Step 1*: Inform yourself about the Alberta tar sands, the key issues and struggles involved, and their implications for our environmental, energy and social future. This book and other publications can serve as a primer for this purpose. For more up-to-date information on what is happening, check out the website www.tarsandswatch.org.
- *Step 2*: Identify the issue or issues involved in the tar sands struggles that speak to you personally. It may be global warming, peak oil or water depletion; energy security, manufacturing jobs, or military links; boreal forests, First Nations, or community struggles; or some other issue or theme. Whatever the issue, it's important that you "take it personally." The issue you choose is your entry point into the struggle.
- *Step 3*: Connect the dots between the issue you have chosen and other issues or concerns in your own life and the lives of your family and friends. Links also need to be made between your issue and the other major tar sands issues we have been discussing. By doing this you will be better able to cultivate your commitment to the tar sands struggle for the long haul.
- *Step 4*: Join a group or organization that is engaged in creative action on tar sands issues, an environmental, union, church or public interest group. If no such group exists in your community, then consider the possibility of identifying others who share similar concerns and organizing a tar sands action group,

which could perhaps be affiliated with another organization already engaged in the movement.

• *Step 5*: Choose one or more of the resistance fronts
outlined above as the focal point of your action on
the issues you selected in step 2. Draft a campaign
strategy for action on your issue and present it to your
group. Based on the discussion by your group, rework
your campaign proposal. In doing so, you will develop and hone your skills as an activist.

This five-step process can be helpful in encouraging more people to become actively engaged in the resistance movement
against the tar sands. Active participation in this resistance movement may be no ordinary experience. Indeed, it could be
participation in a unique moment in history. Not only is the tar
sands mega-machine potentially the single biggest energy enterprise of its kind in the world today, causing untold ecological and
social damage to get at the largest known hydrocarbon deposit
deep below the earth's surface, but the resistance to it may also
contain the ingredients for the making of a new social movement. In his book *Blessed Unrest*, Paul Hawken describes the
convergence of three movements taking place today — social justice, environmental sustainability, and Aboriginal peoples — into
one enlarged social movement with the potential to transform
the relationship between the human species and the rest of
nature, for the sake of the planet.[345] As Hawken puts it, this new
movement "has Red and Green arms and a spiritual hub. The
heart and center is indigenous culture."[346]

CHAPTER 8

Dream Change

Building resistance against the tar sands mega-machine alone is not enough to overcome the kinds of environmental, economic and social challenges on the horizon. It is not sufficient to stop the development of tar sands without, at the same time, developing a strategy for the transition to an alternative energy and environmental future. We are, after all, living in a critically important historical moment for the future of the planet. Our industrial civilization and its accompanying economic system have come to the point where it is pushing the ecological limits of the planet itself. Our model of industrialization, based on the plundering of natural resources to achieve relentless economic growth, has become unsustainable. So, too, has the so-called American dream, whose promise of happiness in the form of an increasingly affluent way of life based on the accumulation of wealth and power, remains one of the system's prime motivators. In short, the time has come to start changing the dream of what constitutes the good life.

Changing the dream will no doubt take on different meanings for different people in this country and around the world. But, essentially, it entails developing a sustainable future based on new models of economy and society that are designed to be in harmony with nature. For industrialized countries this calls for bold transformations. At this point in time, the tar sands megamachine stands as a juggernaut in the way of making the transformation towards a sustainable future. In Canada, therefore, how we deal with the tar sands is central to whether we can successfully make the transition needed to a more sustainable economic and social model. For the moment, our task is not to lay out a blueprint of a sustainable economic future but to begin the process of changing the dream and charting the transition.

Sustainability Challenge

The notion that industrial civilization, as we know it, may well be on the verge of collapse, is not something that has been pulled out of thin air. It is based on historical studies by scholars who have examined the causes of the rise and fall of societies though the ages. The most recent is Jared Diamond's best-selling book *Collapse: How Societies Choose to Fail or Succeed*. After surveying the economic and social practices of dozens of ancient and modern societies, Diamond shows that *collapse* is the destiny of societies that ignore ecological limits and the constraints of available natural resources.[347] The practical lesson, concludes Diamond, is that societies that succeed are those that plan and develop their economies taking into account the limits set by available natural resources. Similarly, Joseph Tainter's earlier classic study, *The Collapse of Complex Societies*, demonstrates that complex societies frequently *collapse* because they outgrow the available energy from natural sources. As societies become more and more complex, the returns on energy extracted from the environment to run their systems tend to decline steadily.[348]

The signs of collapse in our industrialized civilization today are perhaps best captured by Jerry Mander's term "the triple crisis," which refers to the dominant ecological threats that are now converging, that will affect the future of the planet and our species.[349] As discussed at some length in Chapter 2, this triple threat consists of (1) the exponential increase of climate chaos due to the heating of the planet caused by the buildup of greenhouse gases; (2) the peaking of oil, natural gas, coal and other fossil fuels, thereby ushering in the end of cheap energy; and (3) the depletion of other natural resources worldwide, including fresh water, forests, fish, wildlife, biodiversity, coral reefs and fertile soils. The convergence of all three of these threats, says Mander, is "deadly." Not only are all three being caused by the operations of contemporary industrial society, but they were inevitable given the underlying values and priorities of that society.

According to Mander, the "triple crisis" itself is primarily generated by a paradigm that places rapid and expanding economic growth at the centre of global, national and institutional aspirations.[350] The driving force behind this paradigm is the pursuit of increasing corporate and individual wealth. To feed this growth, there is a planet-wide race to exploit natural resources and an uncontrolled use of fossil fuels. In this prevailing model of resource development, there is a profound and systemic disregard for the planetary limits on natural resources, consumption rates and waste generation. Around the world today, says Mander, this model is reinforced by the proliferation of "commodity-oriented economic systems" that depend on the idea that consumerism is fundamental to individual happiness.[351] For these reasons, the unfolding of the "triple crisis" will provoke dramatic shifts in all of the operating assumptions of our industrial society. In short, response to the crisis involves making fundamental changes to the paradigm and the dream that motivates it.

From our vantage point here in Canada, the Alberta tar sands provides a dramatic illustration of the triple crisis unfolding before our eyes. First, the rapid, uncontrolled development of the tar sands not only generates three times the greenhouse gas emissions that conventional oil production does, but the total amount of these emissions will multiply with the doubling, tripling and perhaps quadrupling of current production levels forecast for 2015. At this rate, the tar sands are already expected to account for more than half of this country's growth in carbon emissions over the next twenty years, making Canada a major net contributor to global warming. Second, while the Alberta tar sands possess considerable potential for crude oil production, they could not begin to compensate for the rate of depletion under peak oil conditions. What's more, the net energy output of the crude produced by the tar sands is comparatively low given the amount of energy used to produce it. And instead of reducing our society's oil addiction, the rush to develop the tar sands simply perpetuates it. Just as a drug addict becomes habituated to the drug, by becoming more dependent on tar sands crude, we are getting less bang for our oil buck. Third, the strip-mining operations used to produce the crude from the tar sands are destroying boreal forests in a region that is greater than the size of the Maritime provinces combined. And both the mining and in-situ processes for extracting the bitumen are depleting and contaminating freshwater systems throughout northern Alberta and the southern Northwest Territories.

The Alberta tar sands industry is not only a microcosm of the triple crisis taking place globally. It is also a major generator of the triple threat on this continent. Although mobilizing public resistance to the tar sands in order to prevent further acceleration towards the triple crisis is imperative, so too are campaign strategies aimed at leveraging the tar sands for the transition to a sustainable future. If a national government in Canada were elected on a mandate to make the necessary transition, then a plan of action for the tar sands would be required to enable it. Any plan of action designed to phase out crude oil production in the tar sands, for example, could be carried out for the strategic purpose of gradually reducing consumption rates of oil in both Canada and the US. Corresponding to the production phase-out strategy, a plan for export quotas on crude oil to the US on a scale of increasing reductions over a five- to ten-year period, could be used as leverage to wean the US off of its dependence on foreign oil. Similarly, a conservation plan to reduce consumption of oil and switch to renewable energy alternatives in Canada could be implemented in conjunction with the phase-out plan.

The phasing out of tar sands production needs to be managed in such a way as to facilitate the transition to a sustainable energy and environment future. In order to make a plan to do this, however, it's important to first come to grips with the notion of sustainability. Generally speaking, a sustainable society is one that functions as part of a stable ecosystem. The concept of sustainability was, and is, an integral part of the world views, cultures and traditions of many indigenous peoples. For example, the constitution of the Six Nations of the Iroquois Confederacy, namely, the *Gayanashagowa* or Great Law of Peace, embodies the notion of sustainability.[352] More recently, the idea of sustainability has been revived and given more widespread attention internationally through the 1987 Bruntland Report of the World Commission on Environment and Development. Here, the term "sustainable development," defined as development that "meets the needs of the present generation without compromising the ability of future generations to meet their own needs," was coined. And, in 1992, UBC ecologist William Rees and his colleagues devised a tool for evaluating whether our resource development is sustainable or unsustainable by calculating the ecological footprint of a given human population, that is, the amount of land and water required to provide the resources needed to support it and absorb its wastes, using existing technologies.[353]

While the concepts of "sustainable development" and "ecological footprint" have been influential, says Richard Heinberg, they do not measure up to the challenges of the triple crisis we face today. In his latest book, *Peak Everything: Waking Up to a Century of Declines*, Heinberg proposed a series of axioms about "sustainability." In developing these axioms, Heinberg has drawn on the scientific works of others with the intention of formulating propositions that are both publicly accessible and practical. Below are Heinberg's axioms reformulated for application to some of the key issues we have been discussing about the tar sands.[354]

1. *A society that continues to use critical resources unsustainably will collapse.* A society could choose to change its resource development and consumption patterns rather than collapse, either by consciously deciding not to use critical resources unsustainably or by finding alternatives (while recognizing that any alternative is also bound to be finite). This is the challenge that Canada faces now. A choice has been made to develop, on a massive scale, dirty crude oil from the tar sands as a replacement for conventional oil production. But tar sands crude is a comparatively inferior fuel product because it has less energy-output than conventional oil and it requires more energy-input to produce.[355] Moreover, as noted before, current plans to produce and export crude from the tar sands on an ever-increasing basis serves to perpetuate the US's addiction to oil. As a result, the choice to develop the tar sands' crude resources of the tar sands unsustainably is a false solution to peak oil and will eventually lead to a societal collapse.

2. *Current and projected rates of population growth and resource consumption can not be sustained.* Here, the focus is on the consumption of resources because this is a quantifiable way of

determining the survival of a society in the long run.[356] Resource consumption is also a determinant that can be controlled. In the case of petroleum, it should be clear that significant reductions in oil consumption are imperative, especially in the light of peak oil conditions. But, oil consumption rates in the US, Canada's number one customer, are rising despite the spike in prices at the gas pump. The call for a five-fold increase in tar sands crude production for export to the US is designed, in large measure, to compensate for this increasing consumption rate. Nor has Canada adopted conservation measures in response to its dwindling natural gas and conventional oil reserves. In other words, the development of the tar sands on the massive scale now planned will do nothing to reduce consumption rates and will perpetuate the addiction to oil in both countries.

3. *The rate of declining consumption must be greater than or equal to the depletion rate of the non-renewable resource itself.* In keeping with the previous axiom, relentless consumption of any non-renewable resource is unsustainable. But, if the rate of consumption were to decline faster than or equal to the rate of depletion, then the use of the resource would be sustainable, that is, use would become negligible before the resource was depleted. To facilitate this formula for the reduction in the consumption of oil on an international level, the Oil Depletion Protocol, which calls on oil-importing countries to reduce their imports by the world depletion rate of oil (estimated to be around 2.5 percent per year) and oil-producing countries to reduce their production by their own national depletion rates, has been proposed.[357] If this formula were followed, major advances could be made towards sustainability in oil use. Yet, Canada, spurred on by the tar sands boom, is moving in the opposite direction in terms of both oil consumption and production.

4. *The rate of renewable resource use must be less than or equal to the rate of natural replenishment.* Renewable resources can also be

depleted. Clear-cut logging of forests and overfishing of lakes and oceans can result in the exhaustion of those resources. So, too, fresh water, which is replenished through the water cycle (i.e., whereby precipitation falls to the earth, seeps into the ground, and evaporates into the atmosphere to return again in the form of precipitation), can be depleted. As we saw in Chapter 2, the water cycle is often damaged or disrupted by industrial production, resulting in the drying up, or contamination of aquifers, groundwater systems, streams, lakes and rivers. The tar sands production processes provide a prime example of how fresh water is used and abused at a rate that is greater than the rate of natural replenishment. The use of three to five barrels of water to produce one barrel of crude oil from the tar sands has led to a serious lowering of the water levels in the Athabasca River and connecting river systems. At the same time, the toxic lakes of contaminated wastewater from the production plants leach into both the groundwater systems and the river, to the detriment of the people and wildlife who depend on these sources for their water.

   5. *Contaminating substances introduced into the environment must be minimized and rendered harmless to biosphere functions.* This axiom refers, of course, to a wide range of chemical, radiological, and bacteriological substances from industrial processes and other human activities. In cases where the viability of ecosystems is threatened by pollution generated by the extraction and consumption of non-renewable resources over a period of time, then the "reduction in the rates of extraction and consumption of those resources may need to occur at a rate greater than the rate of depletion."[358] When applied to the production processes of the tar sands, this axiom and its corollary have major implications in relation to both greenhouse gas emissions and the contamination of fresh water systems alone. As we have seen, the heating of the planet is exacerbated by the multiple greenhouse gas emissions from tar sands production, and the boreal ecosystem in the

northwest region of Canada is seriously threatened by the toxic lakes generated by the tar sands industry. As a result, substantial reductions in, if not a complete curtailment of, the production and consumption of tar sands crude are needed to ensure sustainability.

The extent to which Heinberg's system is understood and applied will determine the degree to which a society's production and consumption of both non-renewable and renewable resources is rendered sustainable or unsustainable. Failure to develop policies and practices that are consistent with these five axioms will surely accelerate unsustainable practices, thereby setting the stage for societal collapse. What this formula provides, in effect, is a test of sustainability for resource production and consumption. Applied to plans for the expansion of crude oil development from the tar sands over the next two decades, the result is failure on all five counts. If acceleration of the triple crisis and the impending societal collapse is to be averted, then a plan needs to be developed now to apply all five of these axioms to the phase-out of tar sands production and consumption.

One axiom missing from this formula has to do with *social and economic equity*. Certainly, when it comes to resource depletion and the impacts of rapacious resource production and consumption, the poor and marginalized sectors of society are usually the ones who bear the brunt. In the case of the tar sands, for example, the global warming and water depletion and contamination effects of rapid resource production and consumption have a direct and immediate effect on the Aboriginal peoples who occupy land downstream, making this an issue of climate justice. Similarly, the trickle-up mechanisms for wealth distribution in Alberta whereby the oil boom results in increasing pockets of poverty, or the lack of a national energy strategy that leaves Quebec and Atlantic Canada vulnerable to worldwide oil supply shocks, underscore the imperative for social and economic equity. Although agreeing with these claims. Heinberg points out that the formula for sustainability is aimed at outlining the conditions necessary for maintaining a society over time and averting societal collapse, not the creation of a just society.[359] Even so, achieving social and economic equity would need to be part of the package of solutions to the tar sands challenge.

While the test for sustainability has moral implications, it is, first and foremost, intended to be a recipe for human and planetary survival. As formulated here, it is primarily designed to enable nations and communities to make choices about sustainable alternatives in order to avert the collapse of civilization in the wake of the triple crisis. If the people of Canada fail to apply the test of sustainability to public policy and management decisions affecting the future development of the Alberta tar sands, we do so at our own peril.

Energy Transitions
For Canada, therefore, dealing with the tar sands will be the key to any strategy for making the transition to a sustainable future. But, the sustainability challenge itself cannot be addressed in a vacuum. In other words, simply applying the sustainability test to the tar sands and developing a corresponding strategy for managing the phase-out of the industry alone would not be sufficient. Instead, a viable tar sands plan must be framed in the context of a new energy and environment strategy for the country. In the light of the triple crisis, the question for us is what kind of energy and environment strategy is needed for Canada to make the transition from an unsustainable to a sustainable future. To be effective, such a strategy would have to be made-in-Canada, keeping in mind both our continental and global responsibilities. By framing the sustainability challenge along these lines, the stage is set for developing a viable plan of action on the tar sands.

In so doing, the sustainability challenge needs to be taken up in relation to the goal of achieving energy security. As we saw in Chapter 4, the Bush–Cheney administration was building on the foundations laid by previous US governments when it declared energy, and oil in particular, to be a top issue of national security. Under this doctrine of national security, the US can use its military might to secure control over energy supplies around the world, including Canada if necessary. In times of peak oil, when worldwide demand for oil is increasing while supplies of conventional oil are diminishing, public anxieties about energy security are bound to intensify. In turn, these anxieties can also be exploited by a politics of fear to make people believe that military action is necessary to ensure energy security. However, in times of both peak oil and climate change, our notions of energy security need to be reframed. We can no longer cling to an understanding of energy security that is primarily based on fossil fuels. There can be no energy security without environmental security: the two must go hand-in-hand. Achieving energy security, therefore, implies a transition towards a renewable energy future. In short, this is the sustainability challenge that Canada needs to take up in developing a strategy for energy security in this country and making decisions about what to do with the tar sands.

To develop an energy security strategy along these lines, however, requires taking deliberate steps to stop being an energy satellite of the United States. The surrender of control over our energy resources must first be reversed. This can only be done by a change in political will on the part of the Canadian government to reassert national sovereignty over this country's energy resources, particularly oil and natural gas. It is only by regaining control over the production, consumption and export of our petroleum resources that we will have any chance of making a transition to a sustainable energy future. As long as decisions over Canada's energy resources are left in the hands of the US market and transnational oil companies, the chances of making a transition to alternative energy and environmental strategies are next to zero. What's more, by taking back control over our energy resources, Canada would be in a position to assist the US in weaning itself off of its addiction to oil. If framed in terms of making a transition to an alternative and sustainable energy and environment future, Canada's exercise of energy sovereignty could have positive environmental outcomes for the continent and the planet as a whole.

Yet, to move in this direction would require a significant shift of the locus of energy decision-making from the market to the state, a shift in keeping with trends developing all over the world. Elsewhere, the power to control global oil supplies and distribution has been shifting from the US and the big oil transnationals to oil-producing countries with national energy strategies. Today, state-owned oil companies hold 77 percent of the world's oil supplies. A wave of resource nationalism has swept through the petroleum-producing countries of Latin America, for example, manifested by nations ranging from Venezuela, Bolivia and Ecuador to Brazil and Argentina. In some cases, long-term state-to-state contracts are replacing spot markets for oil. For Canada, this shift would not have to mean a return of the National Energy Program as it was conceived almost three decades ago. But it would require developing a plan for securing democratic national control over the oil industry, including the tar sands, with regards to its basic operations and decision-making, including matters of ownership and investment. Unless Canada is willing and able to make bold moves in this direction, the chances of effectively confronting the triple crisis are minimal.

A made-in-Canada energy strategy could not be pursued effectively by the federal government without collaboration with the provinces. If possible, the best way to proceed is through federal–provincial partnership agreements on energy and environmental priorities and the strategies to attain them. The following is a ten-point platform of measures, drawn from several sources, which largely, but not exclusively, pertain to the petroleum sector.[360]

- *Energy Conservation*: A bold energy conservation plan is needed which would include mandatory energy audits and efficiency programs for all industries, businesses, homes, and community and government facilities. Federal and provincial incentives should be provided for insulation upgrades, air sealing, motor replacements, refrigeration improvements and related conservation measures. Manufacturing and other industrial production processes should be assessed for energy waste and incentives introduced for upgrading energy efficiencies, including combined heat and power generation. As well, before Canada can actually exercise its energy sovereignty by reducing oil and gas exports, it must demonstrate its determination to conserve energy.

- *Energy Reserves*: Since virtually every other oil-producing nation in the world has a strategic petroleum reserve in place, it is high time that Canada did the same. To begin, Ottawa should restore the twenty-five-year rule of proven oil supplies before exports are permitted and Alberta should reinstate its thirty-year proven supply rule before allowing exports. In addition, steps should be taken to establish storage sites for strategic petroleum reserves in Manitoba, Ontario, Quebec and the Atlantic provinces. For example, the salt caverns of Lambton County in western Ontario are believed to have the capacity to store up to thirty-one million barrels of oil, which could be used for emergency needs.[361] Given the prospect that oil shocks are likely to become more frequent, it's important that petroleum reserve sites be located in these and other strategic locations across the country.

• *Energy Security*: Immediate measures need to be taken to ensure that eastern Canada has access to adequate supplies of domestic oil and is not left vulnerable to oil shocks due to disruptions in world markets. To this end, the direction of the Montreal to Sarnia pipeline must be both reversed and expanded so that oil from Alberta and Saskatchewan can flow to meet energy needs in Quebec and Ontario. At the same time, oil from the Hibernia plants in Newfoundland and natural gas from Nova Scotia need to be redirected from export markets in the New England states to meet the energy needs of Atlantic Canada. In effect, there needs to be a Canada First policy whereby our oil and natural gas supplies are made available to serve the energy needs and security of our own population before being exported.[362]

• *Energy Export Quotas*: In keeping with the above requirements for energy conservation, reserves and security, a program of export quotas on oil and natural gas should be introduced. Issuing export quotas comes under the jurisdiction of the federal government and would require changes in NAFTA (see below). The quotas could be introduced on a progressively declining scale over a five- to ten-year period. The criteria for the declining quotas would be set in accordance with the criteria established for the proposed Oil Depletion Protocol, calling on producing countries to reduce their exports at the rate of depletion of their own conventional oil reserves. These export quotas would also be designed to serve as a policy tool for encouraging the US to wean itself off of its addiction to oil.

- *Energy Sovereignty*: In order to impose export quotas on oil and natural gas to the US, Canada would need to reassert its sovereignty over energy by negotiating changes in NAFTA. The people of Canada cannot be expected to make drastic reductions in their energy use if most of the oil and natural gas saved is being exported to the US. In particular, Canada must take advantage of the recent opportunities, initiated in the US presidential race, to re-open negotiations on NAFTA in order to remove the proportionality clause, which essentially makes energy exports from this country to the US compulsory. As noted before, Mexico insisted on and was granted an exemption from the proportionality clause in negotiating its own participation in NAFTA. At the very least, Canada should negotiate an exemption if possible; if it is not, we should invoke the abrogation clause and get out of NAFTA altogether.

- *Energy Ownership*: To regain Canadian control over energy along these lines, measures should be taken to either nationalize or provincialize strategic components of the oil and gas industry. In contrast to for-profit oil corporations, who are legally obligated to provide a return on investments to their shareholders, crown corporations are by definition designed to fulfill public policy goals such as the energy conservation and equity objectives outlined above. Moreover, the remaining private Canadian companies operating in the Alberta oil patch are ripe for further foreign takeover, making energy decisions even more vulnerable to external controlling inter-

ests. Public opinion polls show support for nationalizing sectors of the petroleum industry. According to
a 2005 Leger poll, 51 percent overall and 60 percent of
the youth in this country favour nationalizing the oil
companies operating in Canada.[363]

• *Energy Alternatives*: Reducing consumption of oil,
natural gas and other fossil fuels through conservation measures must be complemented by the
development of energy alternatives. As a society, we
currently rely on fossil fuels for over 80 percent of our
energy needs. While a variety of options exist, including hydropower, wind power, geothermal power, tidal
power, nuclear power, solar power and agro-fuels
such as ethanol, it's important to recognize that they
all have net energy ratios that are less than that of
conventional oil. Nevertheless, investment in renewable energy technologies is critical for developing an
alternative energy strategy, while recognizing that a
number of technical challenges remain to be resolved
before these alternative sources can be scaled up to
the point where they can begin to replace fossil
fuels.[364]

• *Energy Products*: Any responsible energy policy needs
to put an end to the direct export of raw and upgraded bitumen from the tar sands to the United States. At
the very least, priority must be put on upgrading and
refining bitumen here for multiple uses, and also on
the secondary manufacturing of oil derivatives. However, it is imperative that these forms of value-added
production be carried out in a way that is environmentally sustainable. A strategy for value-added
production of oil will certainly create more stable
employment and can also provide opportunities for

preserving the resource over a longer period of time. As well, such a strategy needs to make distinctions between the manufacturing of oil-based products that are necessary for, and compatible with, a sustainable future and those that are not.

• *Energy Emissions*: Obviously cutting carbon emissions must be a top priority for any new energy strategy in Canada. When it comes to the climate change challenge, the reduction of industrial carbon emissions based on intensity targets, which allows industries such as the tar sands to increase their volume of emissions as they expand their production four- or five-fold in the coming years, is not the way to go. Instead, absolute caps on industrial emissions should be made mandatory, and the caps should be progressively lowered in order to meet Canada's targets under the Kyoto Accord (at the very least). Failure to comply with these emission caps should result in heavy fines, shutdowns, and even public takeover of industries. At the same time, carbon emission and fuel efficiency standards for all vehicles on a par with the more stringent ones of Europe and Japan need to be mandated. If properly designed, a carbon tax that recycles revenues to provide incentives for reducing greenhouse gas emissions could prove to be the most efficient means for accomplishing these objectives.

• *Energy Revenues*: The government royalty regimes for the oil and gas industry need to be overhauled. In comparison to Norway, which collects three to six times more in royalties per barrel of oil, the Alberta regime effectively gives away billions of dollars in "unearned" or "excess" profits to the petroleum industry. The main beneficiaries of these revenues should be the owners of the resource, namely, the citizens of energy-producing provinces and the First Nations whose land was taken in the first place. At the same time, federal–provincial agreements need to be worked out for allocating a portion of resource revenues for investment in the development of renewable energy technologies, effective conservation and environmental strategies, just transitions for workers in the tar sands and other fossil fuel industries to jobs in green industries, and to ensure that poor and marginalized peoples are not penalized by higher fuel costs in a post-carbon society.

This ten-point energy strategy is only a sampling of what needs to be done to make the transition from an unsustainable to a more sustainable future. For action purposes, what an energy strategy like this does is establish an alternative framework in which to situate the call for a moratorium on the tar sands. Clearly, control of the development of the tar sands would be pivotal to carrying out a plan such as this. Despite the immense challenge of the triple crisis, however, imposing an immediate moratorium on all existing and proposed tar sands plants would likely be politically unacceptable at the moment. What may be more politically feasible is the imposition of an immediate moratorium on all new tar sands projects. While this would still allow the existing megaprojects to continue doing enormous damage in terms of global warming, water depletion, and social upheaval, the energy export quotas, energy security, carbon emissions targets and other measures outlined above, if adopted, would serve to substantially curb the operations of existing tar sands plants as well. What's more, it would set the stage for a timed phase-out of the tar sands industry itself, in a way that would provide for a just transition of workers to alternative industries.

In the meantime, there are interim measures within federal jurisdiction that could be taken now that would have the effect of slowing down the rapacious pace of tar sands development and setting the stage for a moratorium. In its 2008 Alternative Federal Budget, for example, the Canadian Centre for Policy Alternatives proposed the following steps be taken in order to manage the current resource boom: (1) restore the federal corporate income tax rate to 28 percent for the oil and gas industry, thereby raising approximately 1.75 billion dollars a year in new federal revenues; (2) implement a new federal environmental review and approval process for new mining and oil sands investments, aimed at keeping aggregate greenhouse gas emissions from those projects in line with federal emissions reduction targets; (3) strengthen the regulatory powers of the National Energy Board so that it can make approval of permits for the export of oil and natural gas, and the construction of new export-oriented pipelines contingent on security of supply for Canadian consumers; (4) reform Investment Canada legislation to establish a transparent and binding net benefit test for foreign acquisitions of Canadian-owned companies; and (5) impose a 25-percent excess profits surtax on petroleum production, to be integrated with the new federal carbon-pricing strategy.[365]

The Alternative Federal Budget also contains concrete proposals for a carbon tax. At a rate of 30 dollars per tonne, this carbon tax would initially generate 7 billion dollars a year. The revenues collected from this carbon tax would be allocated to programs for the transition to a greener economy. One set of investment priorities would be renewable energy, energy efficiency and retrofit programs and public transit. A second set of priorities would involve a green energy tax refund to compensate low- to middle-income Canadians for major increases in costs created by the tax on carbon emissions. A third set of priorities would be a just

transition strategy for workers and communities affected by the
phase-out of industries that are heavy carbon emitters (e.g., tar
sands plants) along with a jobs investment plan for the development of new and greener industries.[366] Indeed, these and other
measures in the 2008 Alternative Federal Budget provide a fiscal
transition plan for moving towards a more sustainable future.

Like any platform, these energy and environment strategy proposals need to be further refined through public discussion and
debate. While our focus here has been mainly on a strategy for
the petroleum sector, this strategy needs to be coordinated with
strategies for other energy sectors, such as a moratorium on new
coal mines until carbon storage technologies are assured, and the
strategic upgrading of Canada's electricity grid to make it more
sustainable, and focused on east to west, rather than north to
south. The strategy would also need to be integrated with other
plans for making a transition to a post-carbon society. These
include the redesigning of our cities, which have been organized
around the automobile and road building, into liveable and
workable neighbourhoods where people can walk more and travel less; the rebuilding and upgrading of public transit systems to
provide vastly improved transportation services, both within and
between cities and towns, at affordable prices so they can compete successfully with auto travel; and the promoting of
transition towns in which local residents, communities and businesses work together in making a coordinated transition from
fossil fuel dependence to a renewable local economy.[367]

The implementation of an energy strategy along these lines, as
well as other plans for making the transition to a post-carbon
future, would require substantial intervention by the state. This
does not mean there would be no role for the market. On the
contrary, the market has a role but it would not be the central one
it has now. The massive changes required in a relatively short
period of time — reframing national goals, reallocating capital,

regulating industries, reinstituting export quotas, redesigning cities — require a degree of economic planning and organization that can only be carried out by the governments. But simply returning to the old model of centralized economic planning by state-owned institutions will certainly not suffice. The tools of governments, including public ownership and control of strategic sectors, will be necessary to measure up to the challenge of the triple crisis. To be effective, however, the way these tools are used will have to become much more democratic and participatory. Unless the will of the people is mobilized, the chance of making the transition required is very slim.

Meanwhile, a made-in-Canada platform for an alternative energy and environment future like this may well be in tune with the winds of change that appear to be sweeping across the US these days. While, at this writing, it remains to be seen who will win the 2008 race for the White House, there can be little doubt that Senator Barack Obama has made history with his bold and vibrant message for change, including his New Energy for America program. The main goal of Obama's energy plan is to break America's dependence on foreign oil by breaking America's addiction to oil as its prime source of energy. The long-term objective of Obama's plan is to reduce domestic oil consumption in the US by more than 7.6 million barrels a day by 2025, which is more than a third of current consumption rates and more than a quarter of the rates that had been projected for 2025. To accomplish this goal, an Obama administration would invest $150 billion over the next ten years in alternative sources of energy (for example, wind, solar, hydrogen, bio-mass) and the commercialization of plug-in hybrid vehicles, low-emission coal plants and a new digital electricity grid. To pay for these new energy programs, Obama intends to impose a tax on the windfall profits of the oil companies. These revenues would be used in the short term to provide tax rebates for people living in cold weather

states to help them pay their heating bills and for tax credits to
people who purchase advanced energy-efficient vehicles. And, to
ensure that more energy-efficient vehicles are made in the US,
Obama intends to invest several billions of dollars in the US auto
industry. Although there are many hurdles to be crossed before
such an energy program becomes a reality in the US, if implemented, it would create a more favourable climate for Canada to
enact the kind of alternative energy plan outlined in the previous
pages.

To motivate people to make the transition to a sustainable
energy future, however, it's important not to create false illusions.
There is no silver bullet to resolve the triple crisis. The energy that
would have been generated by the full-scale production of crude
oil from the tar sands will not simply be replaced by renewable
energy sources. In fact, there is no known non-petroleum energy
source that has as high a net energy ratio as conventional oil.
While peak oil predictions are still controversial, once these conditions set in and conventional oil reserves decline, people will be
compelled to depend more on non-petroleum energy sources. In
doing so, we will have to learn to "power-down. For the most
part, the alternative forms of energy simply do not have the
power-generating capacity that conventional oil has. But, as sustainable energy specialist Jack Santa-Barbara points out, North
Americans on average consume three times the amount of energy required for optimum levels of well-being. Studies show, says
Santa-Barbara, that while optimal well-being requires up to 110
gigajoules per person per year, North Americans consume about
325 gigajoules annually per person. By powering-down, people in
Canada could reduce their energy consumption by up to two-
thirds, and still have sufficient energy to achieve optimal
well-being.[368]

In order to achieve optimal well-being, says Santa-Barbara, we
need to focus our attention on "the use of energy to meet basic

human needs and differentiate between economic growth and well-being."[369] As an indicator of human well-being, he says, measuring economic growth through the Gross Domestic Product (GDP) is not only inadequate but potentially misleading and dangerous. What are seen as "negatives" from the standpoint of human or environmental well-being (e.g., more cancer cases, expanding food banks, increasing auto accidents) are viewed as "positives" from an economic growth perspective because they contribute to the GDP. What is needed are new quality-of-life indicators, says Santa-Barbara, such as the Genuine Progress Indicator (GPI), which includes environmental and social indicators as well as economic ones. In other words, something like the GPI is a much more accurate way of measuring optimal well-being than the GDP is. By the same token, the amount of energy required for optimal well-being, when measured in terms of the GPI, will be significantly less than that demanded for increasing economic growth through the GDP.

In short, to make the transition to an alternative and sustainable energy future we need a new framework for measuring optimal well-being along with bold transformations, both in our personal lifestyles and in the dominant institutions that shape our economy and society.

The Great Turning

On the opening page of his book *The Great Turning*, David Korten quotes author Joanna Macy:

> Future generations, if there is a livable world for them, will look back at the epochal transition we are making to a life-sustaining society. And they may well call this the time of the Great Turning.[370]

Korten goes on to ask,

by what name will our children and our children's
children call our time? Will they speak in anger
and frustration of the time of the Great Unravel-
ing, when profligate consumption led to an
accelerating wave of collapsing environmental
systems, violent competition for what remained
of the planet's resources, a dramatic dieback of the
human population, and a fragmentation of those
who remained into warring fiefdoms ruled by
ruthless local lords? Or will they look back in joy-
ful celebration on the noble time of the Great
Turning, when their forebears turned crisis into
opportunity, embraced the higher order potential
of their human nature, learned to live in creative
partnership with one another and the living
Earth, and brought forth a new era of human possibility?[371]

Historically speaking, we find ourselves in a "defining
moment," faced with an irrevocable choice between two pathways
— one leading to the Great Unravelling and the other to the
Great Turning. For Korten, choosing the path of the Great Turning means making bold transformations in our society that will
require a mixture of courage, imagination, determination, cre-
ativity and cooperation. The Great Turning, he says, can only be
brought about through three concurrent movements:

- a "cultural turning," a change in social values and a
  deepening spiritual awareness;
- an "economic turning," a major shift in our models of
  production, consumption and wealth distribution; and
- a "political turning," changes in our models of gover-
  nance and distribution of power.[372]

For Canada, the tar sands represent a defining moment involving an irrevocable choice. Will the tar sands struggle unfolding in this country trigger the Great Unravelling or the Great Turning? If it's to be the latter, then we must focus our energies on the cultural, economic and political turnings that need to take place.

Cultural Turning

For the kind of cultural turning that needs to take place, we should start by reviewing the dominant values that motivate our industrial society, and more particularly its prevailing model of resource development. In advanced capitalist societies it is generally held that the prime goal is expanding economic growth, which entails the accumulation of wealth and power, motivated by the pursuit of profit by individuals. In order to accumulate wealth, our industrial society plunders natural resources which has, in turn, generated negative cultural values towards nature and the environment: Since nature is the storehouse of valuable minerals and resources demanded by industrial society, it must be conquered and exploited. As a result, the driving force behind our industrial civilization has not been sustainability, which implies operating in harmony with nature. Instead, it has been the drive to secure power over nature, to dominate and control it.

This clash of culture and nature is evident in the contrasting attitudes about land held by industrial societies and others. In the cultures and traditions of indigenous peoples, the land is understood to be the source of life itself. Land is Life. It is variously called Mother Earth or Pachamama in some parts of Latin America.[373] For most indigenous peoples in Canada and elsewhere around the world, there is a strong spiritual identity with the land which nourishes and preserves life itself. It was commonly understood that the land was "owned" by the Creator or Great Spirit, that it could not be bought or sold, and that the people were to be its guardians and caretakers. This does not mean that the land cannot be developed for resources, but that the model of development must respect, and be in harmony with, nature, and that the land must be returned to future generations in as good condition as it was before the development. By contrast, in the industrial world view land is seen as a means for making money. Like any other commodity, land is viewed as a factor of production that can be bought and sold in the market. From the perspective of industrial society, land is simply another exploitable and alienable resource. It is seemingly endless in its supply of money-making resources and able to constantly withstand rapacious assaults on its ecology.

These divergences in the cultural and spiritual valuing of land, which are present within Canada let alone throughout the world at large, are also dramatically portrayed by looking at contrasting land-use plans for resource development. As we have seen, the oil-bearing lands of the Athabasca, Cold Lake and Peace River regions have been targeted as valuable pieces of real estate by the big oil companies. The land has been carved up like a jigsaw puzzle and doled out to the corporate players by the Alberta government. Based on this land-use plan, the companies that have purchased leases and permits are given the green light to plunder the oil-bearing bitumen that lies below the land's surface. To do so, they strip away the boreal forest, then use huge pieces of machinery to rip up the earth's surface or bore deep holes into it, draining vast volumes of fresh water from surrounding rivers and creating huge toxic ponds, all the while spewing millions of tonnes of greenhouse gases into the atmosphere.

This land-use plan for the Alberta tar sands stands in sharp contrast to the land-use plan of the Dehcho First Nation, located north of Alberta in the Northwest Territories and linked to the tar sands via the mighty Mackenzie River (known as the Great River) which is fed by the increasingly depleted and polluted waters of the Athabasca and Slave River systems.[374] Although the Dehcho lands have not been explored yet for oil and gas reserves, they do harbour other rich minerals, including coal, diamonds and other valuable base metals. As the basis for their land claim negotiations with Canada, the Dehcho developed a series of detailed and comprehensive maps of their land, showing mineral deposits, sacred places, conservation areas, waterways and communities. Before developing their plan, the elders were consulted at great length; they told stories that have come down through many generations and reveal the areas that to be protected and preserved.[375]

As a result, the Dehcho land-use plan divides the land into three categories. The first, and largest, of the areas are the "conservation zones," which include the South Nahanni Watershed National Park Conservancy, designated as areas in which no new non-renewable resource development will be permitted. The second are identified as "development zones," where non-renewable resource development will be permitted under certain conditions based on specified environmental and social criteria. The third are "mixed zones," containing both development and conservation areas. The fact that Ottawa does not accept this alternative land-use plan, objecting in particular to the amount of land put into conservation zones, is one of the main reasons why no land claim agreement has been reached so far with the Dehcho.[376] Self-government is key to the implementation of the Dehcho plan for stewardship of these lands. Yet, as we saw in Chapter 1, the federal government insists that the Dehcho extinguished their rights to this land through the spurious and fraudulent treaty-making process of 1921.

As noted above, the concept of sustainability has been an integral part of indigenous world views, cultures and traditions. Today, the Dehcho plan is cited around the world as a prime example of a land-use model for sustainable resource development. It values the land and creation as Mother Earth, a value deeply rooted in indigenous cultural and spiritual traditions, and thus prohibits the relentless plundering of natural resources such as is seen in the tar sands. At the same time, it allows for the mining and development of non-renewable resources in designated zones while ensuring that the integrity of other, traditional lands is preserved as sacred space and for animals, birds and fish that depend on these lands. The values that underlie the Dehcho land-use plan contain the keys to unlocking the doors to the kind of cultural turning that is required if we are going to make the transition to a sustainable future. If so, the best hope for this kind of turning and transformation of our dominant cultural values may lie in dialogue with the Dehcho and other First Nation peoples in this country.

Economic Turning

The shift in values underlying this kind of cultural turning should serve to help motivate and inspire a transformation of our economic system. The dominant economic model, the accumulation of capital based on the plundering of natural resources, is simply unsustainable. This neo-classical economic model, says economist Herman E. Daly, driven as it is by the growth imperative, is on a collision course with nature itself.[377] By defining "capital" as human-made assets such as goods and services, machines, technology and buildings, says Daly, this economic model denigrates and devalues "natural capital," namely, the resources of the planet that have made all economic activity possible. As a consequence, the natural resources of the planet have been plundered with little or no regard as to their physical or ecological limits. And now, as evidenced by "peak oil," and, in many cases, "peak minerals," and the peaking of other natural resources, the earth's carrying capacities have clearly reached their limits. The "natural capital" of the planet, which has been heavily taxed since the industrial revolution, is now in depletion mode.

With the advance of peak oil and climate change, says Daly, the dominant economic model could be shaken "like a house of

cards" to the point of collapse.[378] For Daly, a sustainable economic model would be what he calls a "steady state economy,"
wherein the economy adapts itself to the dictates of physical reality. According to Daly, this is the classical view of economics
advocated by Adam Smith, David Ricardo and J.S. Mill in the late
eighteenth and early nineteenth centuries. But this classical view
was replaced by neo-classical economics wherein physical reality
(i.e., natural resources and the environment) was made to adapt
to the dictates of the economy. The new physical realities of peak
oil and climate change, giving rise to a shift from cheap to more
expensive energy, could compel a "turning'" towards a steady
state economic model. This transition could be further accelerated by taxing natural resources that are becoming scarce. In other
words, resource use should be taxed more heavily in order to
compel people and institutions to use resources more efficiently.

Recently, Canadian economists have begun to move in a similar direction. The sustainability challenge, says Jim Stanford, "is to
manage interactions between the economy and the environment
so that the economy can continue functioning *without* causing
ongoing degradation to the environment." Stanford goes on to
say: "Sustainability will require weaning the economy from non-
renewable energy and minerals; extensive recycling of materials to
reduce the need for resource extraction; aggressive protection of
natural spaces and habitats; and strict limits on pollution of all
kinds."[379] Others, such as veteran economist Cy Gonick, see the
threats of climate change and peak oil exposing the limits of capitalism itself. "Capitalism is growth driven," says Gonick. "Without
economic growth profits cannot expand and without expanding
profits capitalism would die." A new movement for "eco-social-
ism" is bound to emerge, he argues, because "the capitalist system,
however flexible and resilient it has been over some centuries of
stress, has no acceptable answer to the twin crises of peak energy
and climate change which it itself produced."[380]

From our standpoint, the economic turning towards a more sustainable future requires changing the prevailing model of resource development. Looking elsewhere, we find some radical options emerging. In oil-producing Ecuador, for example, the government has decided to leave a significant portion of its oil in the ground. Some 920 million barrels of heavy oil, located in Yasuní National Park in the Amazon basin, will not be pumped. The reasons are three-fold: the defence of the livelihood, culture and rights of the indigenous peoples in the region; the preservation of the rich and unique biodiversity of this patch of the Amazon forest; and the prevention of the release of the carbon emissions that would be generated by the extraction of this hard-to-get-at heavy oil deposit. By leaving the oil in the ground, Ecuador has turned the dominant economic model around, sacrificing economic growth and profits in favour of the preservation of its natural and cultural heritage, which Ecuadorians contend is irreplaceable.[381]

Clearly, the moratorium on oil production in the Yasuní region of the Amazon has implications for the Alberta tar sands. Indeed, there are also similarities with the campaign to prevent the extraction of crude oil in the wildlife refuge of Alaska and the moratorium on the construction of the Mackenzie Valley Pipeline put in place by the Canadian government in 1977. The big differences, of course, are that Ecuador is a comparatively poor and indebted country with a warm climate, while Alaska, the Yukon and the Northwest Territories are sub-national regions with relatively small populations and harsh weather and terrain. But, the Ecuadorians have calculated the costs and benefits of their action. They know that climate change will have a devastating impact on the poor countries of the global south (e.g., the melting of the Andean glaciers will cause massive water shortages). They also know that the World Bank and International Monetary Fund will respond by making large loans available for dealing with the consequences of climate change, which will, in turn, further increase their indebtedness, and they will be compelled to repay with increasing oil and gas exports.[382] They know not only that the prevailing economic model for resource development is fundamentally flawed, but also that it is unsustainable. Indeed, the cultural turn they have taken shows that they know that developing a new economic model is imperative if a sustainable future is to be realized.

Political Turning
While the example of Ecuador's moratorium on oil development in the Yasuní region of the Amazon illustrates cultural and economic turning that is taking place elsewhere, it is also an example of political turning. Recognizing that preservation of the Yasuní is essential for fighting climate change and protecting the planet, the Ecuadorian government is proposing to rich countries of the global north, via the United Nations, that Ecuador be compensated for its actions, based on a formula that produces an amount that is approximately a third of what the government would receive from the extraction of the oil.[383] The money would be put in a public trust earmarked for paying off the country's external debt, protecting the Yasuní and its indigenous people, and making new environmental and social investments. The government has also taken steps to formalize the moratorium in legislation and establish international safeguards. Furthermore, the Constitution of Ecuador is being rewritten to establish rights for nature, which will make Ecuador the first country in the world to do so.

The possibility of political leadership, such as shown in Ecuador, facilitating a real shift in cultural vision and economic model probably seems far-fetched to many people in Canada. In an era of neo-liberalism, governments have taken a back seat to markets and corporations when it comes to economic and social policy-making, as we have seen in the case of energy policy. The same is true, for the most part, in the case of climate change, let alone peak oil or the rapid depletion of other natural resources, which remain largely ignored. Whether we are talking about federal or provincial government programs to combat climate change, most are advanced market-oriented solutions. While a mix of strategies may be required, it is not at all clear that carbon trading, carbon intensity targets and carbon taxes are sufficient to bring about the changes in patterns of consumption and models of production necessary. If the crisis of peak oil is left mainly to the market and transnational corporations to resolve, there is very likely to be a big investment push for the production of tar sands crude (and for coal and nuclear power), which will further increase greenhouse gas emissions, fresh water depletion and exports of oil to the US. And, in a world marked by growing inequalities, the market-driven escalation in energy prices is having, and will continue to have, horrific consequences for the poor majority. Leaving it to the market alone is simply not sufficient.

Instead, countries such as Canada need to undergo their own political turning, away from a reliance on neo-liberal policies. The threats posed by the triple crisis of global warming, peak oil and depletion of natural resources are matters of life or death for humanity and perhaps even the planet, as we know it. In this historical moment, a new kind of political leadership is called for, one that takes us to a revitalization of the state and a reassertion of its role of acting for the common good. In response to the great upheavals of the Great Depression and World War II, the Canadian government intervened in the economy, to overcome the massive unemployment of the first, and to convert industries to a wartime footing in response to the second. During the 1940s, a model of economic planning was introduced that involved reallocating capital to key industries, rationing of resources, regulating the operations of industries, and instituting price controls. Today, the sheer scale and complexity of the triple crisis, coupled with the speed with which we must respond, demands a concerted and coordinated intervention that can only be marshalled at the national level by the federal government.[384] As noted previously, however, state intervention of this magnitude could not be effectively implemented without, at the same time, democratizing the machinery of the state to allow for greater public or community participation.

What's more, this kind of political turning cannot happen without much closer links being forged between the environmental and labour movement in this country. As emphasized above, the industrial plundering of natural resources that has given rise to the triple crisis, is driven by the accumulation of capital and the prevailing model of production and consumption. The environmental movement needs to recognize that workers and their unions are essential to bringing about the transformation required in our model of industrial production. By the same token, labour unions need to fully understand that the environmental movement is on the cutting edge of social change in this country and that environmentally responsible production is the new front for organized labour. It is here that the struggle for decent jobs and democratic workplaces can be waged, along with the fight for "just transitions" for workers in environmentally and socially destructive industries such as the Alberta tar sands. Building solidarity between these two movements, for the purpose of ending the exploitation of both workers and the environment, could well spark the kind of political turning that needs to take place in this country.

But since politics is about power, there can be no political turning, in the final analysis, without a shift in power and its operational meaning. When it comes to political life in our high-tech, industrialized society, power equals domination. It is the "power" exercised by humanity over nature, by the stronger over the weaker, by the wealthy over the poor, by the technical expert over the ignorant, by the corporate interests over local communities and economies. Here, power also means progress which, in turn, means expansion. The last remaining undeveloped spaces on the planet, such as the boreal forests of the tar sands, are the few places where human power can still be extended in order to advance and conquer, increase control and gain dominion. Once conquered, this power is then manifested in the hierarchical organization of life and activity by those who control capital, technology and resources. But, this understanding of power as domination will not suffice in building a sustainable future. What counts in creating and maintaining sustainability is not power as domination but power as relational. Here, power has to do with the quality of relations between the parts and the whole. For sustainability, power must be exercised to preserve the well-being of the parts within the whole, so that society can co-exist in harmony with nature, both the earth and sky. Our best hopes of retrieving this relational understanding of power may well lie in cultivating a real dialogue with indigenous peoples.

In the end, it all boils down to changing the dream. The long-held American (Canadian, European) dream of an increasingly affluent lifestyle, achieved through the accumulation of wealth and power from the plundering of the earth's natural resources, is simply no longer sustainable. Either we start changing the dream and charting our transition to a sustainable future, or we will surely perish as a people, as a civilization, perhaps as a species, on this planet. For those of us who inhabit this country called Canada, in the upper half of the North American continent, the big test we face is the Alberta tar sands. How we deal with this juggernaut will determine whether or not we are able to make the transition to a sustainable future. So, we must ask ourselves: When our children and grandchildren look back on this moment in history, will they see this as the time of the Great Unravelling or the time of the Great Turning?

APPENDIX

Tar Sands Probe: A Brief Study Guide

Chapter 1. Crude Awakening

This chapter covers the early development of the tar sands with the following subsections: (a) *black gold*: the pioneers who developed the technology to extract oil from the tar sands; (b) *land grabs*: the takeover of indigenous lands, federal and provincial control disputes, and land grants to oil companies; (c) *petro-politics*: the federal-provincial struggle for control over tar sands development and energy policy-making; (d) *energy surrender*: how Canada relinquished control over energy development to the US.

• What impressions did the story of the early development of the tar sands have on you? The sheer magnitude of the tar sands and its oil potential? What about the kinds of strip mining and in-situ methods devised to extract the bitumen?

• How did you respond to the series of land grabs — from indigenous peoples, to provincial and federal governments, to the oil corporations? What does this say about the model of resource development in Canada?

• What comments do you have about the federal-provincial struggle for control over the tar sands? What are some of the positive as well as negative aspects of the National Energy Program? What do we have to learn about the surrender of control over our energy resources through the free trade deal?

Chapter 2. Triple Crisis

This chapter discusses the convergence of three global and potentially deadly crises with implications for the development of the tar sands: (a) *peak oil*: the end of cheap, abundant supplies of oil as global demand outstrips supply; (b) *climate change*: global warming generated by the release of greenhouse gases, mainly from the burning of fossil fuels; (c) *water depletion:* the rapid depletion and contamination of freshwater and other vital natural resources on the planet.

• What comments do you have about the challenges of peak oil, climate change and the depletion of natural resources such as water? Do you see a potentially deadly convergence of these three crises? Is there one that's more important than the others?

• Does this Triple Crisis provide a useful set of lenses through which to look at the development of the tar sands? What do you think are some of the key concerns that should be raised about the tar sands from the perspective of peak oil? Climate change? Water depletion?

• What do these challenges have to say about the development of the tar sands in the future? What can be done to promote public discussion and debate about these challenges and how they relate to the tar sands, both in our communities and throughout Canada?

Chapter 3. Energy Superpower

This chapter concerns the promotion of Canada as "the world's next energy superpower," largely because of the development of the tar sands, with these themes: (a) *corporate players*: the major oil companies and their plants for developing oil from the tar sands, as well as upgraders, pipelines and refinery infrastructure; (b) *economic backbone*: how tar sands development is the centrepiece of a northern vision, as well as the Canadian economy, influencing currency markets, manufacturing exports and balance of trade; (c) *Harper's firewall*: how federal government policies and actions have been designed to protect tar sands industry as the cornerstone of Canada as an en emerging energy superpower.

• What are your impressions of the main corporate players in the tar sands industry and their plans for expanded crude production by 2015? What comments do you have on the magnitude of the planned tar sands development, including upgrader, pipeline and refinery infrastructure? Should our governments be fortifying the tar sands industry so Canada can become an energy superpower?

- Are the tar sands becoming the backbone of the Canadian economy? What happens to our economy when top priority is put on developing and exporting our energy resources? What impact does this have on our manufacturing sector and the rest of the economy? What regions of the country are most directly affected?

- What do you think about Canada becoming an "energy superpower"? Is this the role that Canada should be playing in an age of climate change? In an era of peak oil? What would it mean to be an energy superpower when Canada has surrendered sovereignty over its energy resources to the US?

Chapter 4. Fuelling America

This chapter argues that the prime reason for the rapid development of the tar sands is to provide a secure source of oil supplies for the US. Its sections are (a) *America's oil addiction*: as US oil sources become depleted, the US's strategy is to increase dependency on foreign supplies through imports rather than to reduce domestic oil consumption; (b) *national security*: how energy security has become the number-one priority of national security and how the US military is geared to secure control over foreign oil sources and protect supply routes; (c) *energy satellite*: how Canada's role, through the development of the tar sands, is to be an energy satellite or colony of the US.

- What do you think about Canada being America's fuel pump, now and in the future? What would happen if the US put priority on reducing its oil consumption rather than increasing foreign oil imports and dependency? How far should we go in developing the Alberta tar sands to fuel America's oil addiction?

- What are the implications of the US declaring energy as a priority for its national security interests? What do you think of the US being the "globo-petro cop" by using its military to secure oil sources and supply routes around the world? Could the US military and homeland security forces be used to secure ongoing oil supplies from the Alberta tar sands?

- Do you think Canada has already become an energy satellite or an energy colony of the US? If so, what does this say about our destiny as a people and a nation? How can the development of the tar sands be managed responsibly by Canada if it has surrendered its sovereignty over the country's energy sources?

Chapter 5. Ecological Nightmare

This chapter discusses the multiple forms of environmental damage caused by the massive development of the tar sands for crude exports to the US: (a) *global warming*: the emergence of the tar sands industry as Canada's largest emitter of increasing greenhouse gases or carbon that traps the heat in the atmosphere; (b) *water depletion*: the depletion and contamination of the Athabasca and related river systems due to tar sands production plus the creation of huge toxic ponds; (c) *boreal destruction*: the strip mining of the boreal forest, the muskeg and the wetlands, the best natural storehouse for carbon emissions to be found on the planet.

- What concerns do you have about the tar sands industry potentially generating a four-fold increase in greenhouse gas emissions between 2005 and 2015 due to expanding production? Do you see why intensity targets need to be replaced by absolute targets if Canada is going to effectively reduce its carbon emissions? What action should be taken by the federal government? The Alberta government?
- What concerns need to be raised about how the tar sands industry is depleting freshwater sources in the Athabasca River and related river systems in northern Alberta and the Northwest Territories? What about the large toxic ponds created by the industry and their contamination of nearby groundwater systems? What actions should be taken by the Alberta and federal governments?
- What issues need to be addressed about the strip mining of the boreal forest in this region? What does this have to say about Canada's

responsibilities to protect the Canadian boreal as one of the world's best natural storehouses for greenhouse gas emissions? What actions should be taken by the Alberta and the federal governments?

Chapter 6. Social Upheaval

Chapter 6 discusses the many ways in which the development of the tar sands has become a caldron that is brewing with social conflicts and tensions: (a) *social conflicts*: ranging from the community problems facing Fort McMurray, Upgrader Alley, and Fort Chipewyan, to the growing rich-poor gap in Alberta, the royalty rip-off, and the use of cheap labour; (b) *energy insecurities*: the availability of sufficient oil and natural gas supplies to meet needs in the rest of the country due to increasing exports; (c) *US military links*: the role that Canada has come to play in fuelling America's military operations through the export of tar sands crude.

- What are the hot-button conflicts brewing at the community level in Fort McMurray, Fort Chipewyan or Upgrader Alley? What actions need to be taken by the Alberta government? What other major social tensions are being generated by the growing rich-poor gap, the royalty rip-off, and the use of cheap labour by the tar sands industry? What action should be taken by the Alberta and federal government?

- Do you think people in Quebec, the Atlantic Provinces and parts of Ontario have reason to be concerned about dwindling natural gas supplies in western Canada? What impact would natural gas shortages have on heating of homes and the running of industries? Why does Canada not have a strategic petroleum reserve? What actions need to be taken by the federal government?

- Do you think Canada should be a fuel pump for the US military? What concerns do you have about the tar sands' links to US military operations and what actions need to be taken? Do you see any connections between this and any of the other issues discussed above regarding energy insecurities and social conflicts? If so, what impacts do these have on the rest of Canada? The rest of the continent?

Chapter 7. Resistance Movement

Chapter 7 discusses the building of a social movement of resistance against the massive and rapid development of the tar sands for crude exports to the US: (a) *campaign networks*: the various campaigns that have been organized in Alberta, the rest of Canada and parts of the US; (b) *moratorium platform*: the call for a moratorium on tar sands production as a common platform and basis of unity for this movement of resistance; (c) *resistance building*: strategic opportunities for organizing, broadening and deepening this movement of resistance against the tar sands.

• Do you have any comments on the various campaigns that have been organized in Alberta on the critical issues of the tar sands? In the rest of Canada? In the US? Are you involved in any of these groups yourself? What can be done to make these campaigns more effective?

• What do you think about the call for a moratorium on the tar sands? Should it be a moratorium on the approval of new tar sands projects only or should it cover existing projects as well? Should it be a permanent moratorium leading to a phase-out of the industry itself? If so, what kind of job transition plans would have to be put in place for workers?

• What is your response to the five-step process for becoming active participants in this resistance movement? In which of the 10 fronts for action outlined in the last section of this chapter would you like to be involved? What else could be done to broaden and deepen the resistance?

Chapter 8. Dream Change

This final chapter discusses the need to start changing the dream of the good life based on accumulating wealth and power, including our model of industrialization, and developing strategies for transition to an alternative energy and environmental future: (a) *sustainability challenge*: the reasons why societies collapse when they fail to meet the test of sustainability and how this applies to the development of the tar

sands; (b) *energy transitions*: how to move from the massive development of the tar sands to a made-in-Canada plan of action for transition
to an alternative energy and environmental future; (c) -: how the transition to a sustainable future will require a 'cultural turning,' an
'economic turning,' and a 'political turning.'

• What comments do you have on the sustainability challenge and the
  collapse of societies? Can you see the test of sustainability being
  applied to the development of the tar sands? If so, what judgments
  and conclusions would you draw about the viability of the tar sands
  after applying this test? What changes need to take place in our values
  and priorities as a society and our model of resource development if
  we are to pass the test of sustainability?

• What comments do you have about the proposed 10-point plan of
  action for making energy transitions? Which of the 10 points do you
  feel are most crucial for making the transition to an alternative energy and environmental future? What points do you think are missing?
  Name them. What needs to be done to enable the people of Canada to
  make this kind of energy transition?

• What kind of 'cultural turning' needs to take place in Canada and
  what can be done to stimulate it? What would it mean to develop an
  alternative economic model ('economic turning') for resource development based on new values and priorities ('cultural turning')? And,
  what kind of 'political turning' will be necessary to bring about these
  kinds of changes?

What can be done to promote public discussion and debate about
the development of the tar sands and the critical issues raised in all of
these chapters? How can this be done in ways that communicate and
connect with the key issues and concerns affecting communities
throughout the rest of Canada? What actions can be taken, both in the
short term and in the long term, to make the concerns raised here
about the development of the tar sands a national priority?

Notes

Chapter 1: Crude Awakening

1   For a fuller description of these events, see William Marsden, *Stupid to the Last Drop* (Toronto: Alfred A. Knopf, 2007), pp. 25–28.

2   Cited in David H. Breen, *Alberta's Petroleum Industry and the Conservation Board* (Edmonton: University of Alberta Press, 1993), p. 8.

3   Larry Pratt, *The Tar Sands: Syncrude and the Politics of Oil* (Edmonton: Hurtig Publishers; 1976), p. 34

4   Larry Pratt, *The Tar Sands*, p. 34.

5   See Marsden, *Stupid to the Last Drop*, pp. 29 ff.

6   Marsden, *Stupid to the Last Drop*, p. 30.

7   Pratt, *The Tar Sands*, pp. 38–41.

8   Marsden, *Stupid to the Last Drop*, pp. 30–33.

9   Pratt, *The Tar Sands*, pp. 35–36.

10   For a fuller description of Manley Natland's proposals to use nuclear explosions to get the bitumen flowing in the tar sands, see Marsden, *Stupid to the Last Drop*, pp. 2–5 ff.

11   Cited in Marsden, *Stupid to the Last Drop*, p. 38.

12   Marsden, *Stupid to the Last Drop*, p. 38.

13   Marsden, *Stupid to the Last Drop*, p. 40.

14   For an overview of Aboriginal rights, see Hugh McCullum and Karmel McCullum, *This Land Is Not For Sale* (Toronto: Anglican Book Centre, 1975), chapter 1.

15   Research on Treaty 8 was carried out by priest–historian Fr. Rene Fumoleau and documented in his book, *As Long As This Land Shall Last* (Toronto: McClelland & Stewart, 1974, re-published with new data by University of Calgary Press, 2004).

16   McCullum & McCullum, *This Land Is Not For Sale*, pp. 14–16.

17   Marsden, *Stupid to the Last Drop*, p. 32.

18   See Paul Chastko, *Developing Alberta's Oil Sands* (Calgary: University of Calgary Press, 2004), pp. 35–36.

19   Pratt, *The Tar Sands*, p. 41.

20   Cited in Marsden, *Stupid to the Last Drop*, p. 34.

21   Cited in Pratt, *The Tar Sands*, p. 41. See also pp. 42 ff.

22   Pratt, *The Tar Sands*, pp. 30–31.

23   The details of the permits and leases are outlined in Pratt, *The Tar Sands*, p. 32.

24   Pratt, *The Tar Sands*, p. 43.

25   Pratt, *The Tar Sands*, p. 45.

26   Cited in Pratt, *The Tar Sands*, pp. 44–45.

27   Pratt, *The Tar Sands*, pp. 9–10.

28   See details in Pratt, *The Tar Sands*, pp. 46–47.

29   For description of Peter Lougheed's press conference on the Syncrude deal, see Pratt, *The Tar Sands*, pp. 19–21.

30   For an outline of the Trudeau government's 1973 energy policy see Paul Chastko, *Developing Alberta's Oil Sands* (Calgary: University of Calgary Press; 2004), pp. 152–153.

31   Chastko, *Developing Alberta's Oil Sands*, p. 152. The following two paragraphs are based on Chastko's work.

32   Cited in Chastko, *Developing Alberta's Oil Sands*, pp. 155–157.

33   Pratt, *The Tar Sands*, p. 10.

34   Chastko, *Developing Alberta's Oil Sands*, pp. 172–173.
35   *The National Energy Program* (Ottawa: Minister of Supply and Services, 1980), p. 2.
36   For a fuller description of these NEP components, see Chastko, *Developing Alberta's Oil Sands*, pp. 180–182.
37   Peter Lougheed, *Transcript of Televised Address to Albertans*, October 30, 1980.
38   Lalonde's reaction to Lougheed's televised address as distinct from that of Trudeau, notes Chastko in *Developing Alberta's Oil Sands*, p. 186. See also Geoff White, "Door Open for New Oil Talks," *Calgary Herald*, October 31, 1980.
39   Although the NEP acknowledged that the tar sands industry needed higher prices in order to develop crude oil, the price "need not be as high as the international price." Instead of adopting a world price for tar sands oil, a "New Oil Reference Price" (NORP) would be established for the oil sands and heavy oil projects. See Chastko, *Developing Alberta's Oil Sands*, pp. 180–185.
40   The tar sands industry argued that the new agreement on the tax regime did not take into account the interrelationship between the conventional oil industry and the tar sands industry. Profits made in the conventional industry often go to enable the company's operations in the tar sands. As a result, the doubling of the base rate in the new tar sands regime agreement from 8 percent to 16 percent was viewed as penalizing conventional oil-producing companies and as a consequence the tar sands industry.
41   Chastko, *Developing Alberta's Oil Sands*, pp. 186–195.
42   For the author's description of these events in a previous publication, see Tony Clarke, *Silent Coup: The Big Business Takeover of Canada* (Toronto: James Lorimer & Co., 1997), pp. 22–25.
43   See Stephen Clarkson, *Canada and the Reagan Challenge* (Toronto: James Lorimer & Co., 1985).
44   See Chastko, *Developing Alberta's Oil Sands*, p. 193.
45   For a description of the BCNI strategy and these events, see Clarke, *Silent Coup*, pp. 22–25.
46   See Clarkson, *Canada and the Reagan Challenge*, and Clarke, *Silent Coup*, p. 24.
47   Clarke, *Silent Coup*, pp. 26–27. See also David Langille, "The BCNI and the Canadian State," *Studies in Political Economy* (Autumn, 1997).
48   Clarke, *Silent Coup*, p. 27.
49   For a discussion of the energy chapter and its impacts, see Maude Barlow, *Parcel of Rogues: How Free Trade is Failing Canada* (Toronto: Key Porter Books, 1990), pp. 129–149.
50   Barlow, *Parcel of Rogues*, p. 134.
51   Barlow, *Parcel of Rogues*, p. 132.
52   Peter Lougheed, "The Rape of the National Energy Program Will Never Happen Again," in Earle Gray, ed., *Free Trade, Free Canada* (Woodville, Ontario: Canadian Speeches, 1988).

Chapter 2: A Triple Crisis

53   Jerry Mander, ed., *Manifesto on Global Economic Transitions: Towards a Global Movement for Systemic Change, Economics of Ecological Sustainability, Equity, Sufficiency and Peace*, A project of the International Forum on Globalization, the Institute for Policy Studies, and the Global Project on Economic Transitions (San Francisco: 2007), pp. 1–4.
54   For background on Marion King Hubbert, see Richard Heinberg, *The Party's Over: Oil, War and the Fate of Industrial Societies* (Gabriola Island: New Society

Publishers, 2005), pp. 95–101.

55 Heinberg, *The Party's Over*, pp. 101–110.

56 Robert L. Hirsch, Roger Bezdek and Robert Wendling, "Peaking of World Oil Production: Impacts, Mitigation and Risks," prepared for the United States Department of Energy, February, 2005.

57 Hirsch, Bezdek and Wendling, "Peaking of World Oil," p. 5.

58 See, for example, Planet for Life, "The Problem of Reliable Data," http//planet forlife.com/oilcrisis/oilreserves.html. See also Energy Watch Group, Crude Oil: The Supply Outlook, 2007, http://www.energywatchgroup.de/fileadmin/global/pdf/EWG_Oilreport_102007.pdf.

59 Reuters, "Kuwait oil reserves only half official estimate — Petroleum Intelligence Weekly," January 20, 2006. Cited in Gordon Laxer, *Freezing in the Dark*, a joint report of the Parkland Institute and Polaris Institute, February 2008, p. 13.

60 Cited in Paul Roberts, *The End of Oil* (Boston: Houghton Mifflin; 2004), p. 52.

61 Bassam Fattouh, "Spare Capacity and Oil Price Dynamics," *Middle East Economic Survey*, Vol. XLIX No. 5, January 30, 2006.

62 Bengt Soderbergh, Frederik Robelius, Kjell Alekett, "A Crash Programme Scenario for the Canadian Oil Sands Industry," Energy Policy 35 (2007), pp. 1942–1943.

63 Soderbergh, Robelius and Alekett, "A Crash Programme," p. 1946.

64 The US Geological Survey issued a report on July 24, 2008, saying that there may be up to 90 billion barrels of "technically recoverable" oil in the Arctic Circle. These oil deposits lie under the waters of the continental shelf which are currently claimed by several countries — Canada, Russia, Norway, Denmark and the United States. See Linda Diebel, "Northern Oil Riches Raise Fears," *The Toronto Star*, July 24, 2008, p. A19.

65 Soderbergh, Robelius and Alekett, "A Crash Programme," pp. 1946-7.

66 Cited in Richard Heinberg, *The Party's Over: Oil, War and the Fate of Industrial Societies* (Gabriola Island: New Society Publishers, 2005), pp. 108–110.

67 Matthew R. Simmons, "Gauging The Risks of Peak Oil," ASPO World Conference, October 18, 2007, in Houston, Texas.

68 Cited in Hugh McCullum, *Fuelling Fortress America* (Ottawa: Canadian Centre for Policy Alternatives, the Parkland Institute and the Polaris Institute, 2006), p. 21.

69 Tyler Hamilton, "Economist predicts $1.50 a Litre of Gasoline," The *Toronto Star*, January 4, 2008.

70 Hirsch, Bezdek and Wendling, "Peaking of World Oil," p. 5.

71 Heinberg, *The Party's Over*, pp. 91–92.

72 Securing America's Future Energy, "Oil Shockwave: Oil Crisis Executive Simulation," September 6, 2005, www.secureenergy.org.

73 Laxer, *Freezing in the Dark*, p. 13.

74 See also Al Gore's book, *An Inconvenient Truth: The Planetary Emergency of Global Warming and What We Can Do About It* (Emmaus, Pennsylvania: Rodale Publishers, 2006).

75 For a brief account of these events concerning the climate change challenge, see Cy Gonick, ed., *Energy Security and Climate Change: A Canadian Primer*, Introduction, p. 11ff. Also see Dale Marshall, "Fudging the Numbers: Stephen Harper's Response to Climate Change," in the same volume, pp. 149-160.

76 George Monbiot, *Heat: How to Stop the Planet from Burning* (Toronto: Random House of Canada; 2006), pp. 3–4.

77 Bill Hare and Malte Meinshausen, *How Much Warming Are We Committed To*

and How Much Can Be Avoided, PIK Report 93 (Potsdam Institute for Climate Impact Research, Potsdam, 2004), p. 24.

78 Monbiot, *Heat*, p. 15 ff.

79 Monbiot, *Heat*, p. xvii

80 Monbiot, *Heat*, pp. 6–7 ff.

81 For details see Meteorological Office, "Impacts on Human Systems Due to Temperature Rise, Precipitation Change and Increases in Extreme Events" in *International Symposium on the Stabilization of Greenhouse Gases: Tables of Impacts* (Hadley Centre, Exeter, 2005).

82 Cited in Monbiot, *Heat*, pp. 15–16 ff.

83 Jim Hansen, "The Threat to the Planet," *New York Review of Books*, July 13, 2006, pp. 12–16. Cited in Gonick, ed., *Energy Security and Climate Change*, pp. 11–12.

84 Monbiot, *Heat*, pp. 27–28 ff.

85 For a discussion of the alliances or coalitions of corporations on climate change see David Noble, "The Corporate Climate Coup," in Gonick, ed., *Energy Security and Climate Change*, pp. 135–143.

86 The GCC effectively put climate change issues on hold, says Noble, who cites *Los Angeles Times*, December 7, 1997. See also entries for 'climate change' at http://www.sourcewatch.com and http://www.commoncause.com

87 Nicholas Stern, "Stern review on the Economics of Climate change," 2006, at http://www.hm-treasury.gov.uk/independent_reviews/stern-review-economics -climate_review_economics_climate_change/stern-review_index.cfm.

88 David Suzuki Foundation et al., "Assessment by Canadian Environmental Leaders of the Government's Kyoto Implementation Plan," press release, 2005. http://www.david suzuki.org/files/climate/KyotoLeadersStatement.pdf.

89 For a discussion of the Stephen Harper government record in response to climate change challenge, see Dale Marshall, "Fudging the Numbers," pp. 150–153 and 155–157.

90 Cited in Gonick, ed., *Energy Security and Climate Change*, p. 20.

91 I have authored and co-authored two books and numerous articles on water issues including *Blue Gold* (with Maude Barlow in 2002) and *Inside the Bottle: Exposing the Bottled Water Industry* (2007). This section of Chapter 2 draws on some of this work.

92 Most of Michal Kravčík's work is in the Slovakian language. However, a summary of his work has been published in English entitled *New Theory of Global Warming* published in 2001. For a discussion of Michal Kravčík's studies and their implications see Maude Barlow and Tony Clarke, *Blue Gold: The Battle Against the Corporate Theft of the World's Water* (Toronto: Stoddart Publishers, 2002), pp. 10–12.

93 My calculations based on Kravčík's conclusions. See Barlow and Clarke, *Blue Gold*, p. 11.

94 For description of this water contamination around the world, see Maude Barlow and Tony Clarke, *Blue Gold*, chapters 1 and 2.

95 For more detailed information and analysis of this argument, see John B. Sprague, "Great Wet North? Canada's Myth of Water Abundance," in *Eau Canada: The Future of Canada's Water*, Karen Bakker, ed. (Vancouver: UBC Press, 2007), pp. 23–35.

96 Sprague, "Great Wet North?"

97 Urban Water Council, 2005. National City Water Survey, 2005, Washington, D.C., November 15, 2005. http://www.usmayors.org/74theWinterMeeting/National

CityWaterSurvey2005.pd. For an analysis of this data, see Tony Clarke, "Turning on Canada's Tap?" (long version), Polaris Institute, April 2008, available at www.polarisinstitute.org.

98   The articulation of the looming water crisis in these three regions of the US is drawn directly from my own work. See Tony Clarke, "Turning on Canada's Tap?" (Polaris Institute, April 2008), available at www.polarisinstitute.org.

99   For background on these three water export mega-corridors, see Frederic Lasserre, "Les Projets de Transferts Massifs d'Eau en Amerique du Nord," *Vertigo*, hors-serie no. 1. See also two articles by the author, "Turning on the Tap?" in Bruce Campbell and Ed Finn, eds., *Living with Uncle: Canada–US Relations in an Age of Empire* (Toronto: Lorimer & CCPA, 2006), pp. 94–119, and "Turning on Canada's Tap," Polaris Institute.

100   David Schindler, "The Cumulative Effects of Climate Warming and Other Human Stresses on Canadian Freshwaters in the New Millennium," *Canadian Journal of Fisheries and Aquatic Sciences* 58, pp. 18–29.

101   James Howard Kunstler, *The Long Emergency: Surviving the Converging Catastrophes of the 21st Century* (New York: Atlantic Monthly Press, 2005).

Chapter 3: Energy Superpower

102   For the relevant excerpt from Stephen Harper's speech in London on July 14, 2006, see Harper's Index at http://www.dominionpaper.ca/articles/1491.

103   For a comprehensive analysis of Stephen Harper's use of the term, see Annette Hester, "Canada as the 'Emerging Energy Superpower': Testing the case," Canadian Defence & Foreign Affairs Institute, October 2007.

104   Larry Pratt, *The Tar Sands: Syncrude and the Politics of Oil* (Edmonton: Hurtig Publishers; 1976).

105   Larry Pratt, *The Tar Sands*, pp. 9–10.

106   Larry Pratt, *The Tar Sands*, pp. 19–21.

107   Larry Pratt, *The Tar Sands*, p. 9.

108   Annette Hester and Sidney Weintraub, Canada Chapter, in *Energy Cooperation and Impediments in the Western Hemisphere*. Sidney Weintraub, ed., with Annette Hester and Veronica Prado (Washington, DC: CSIS, 2007), p. 71.

109   The following information on the tar sands projects being developed by the major corporate players was compiled by Richard Girard, lead researcher at the Polaris Institute. The information came from the Government of Alberta's Statistics and Publications, Strategy West Inc., and the websites of the companies involved.

110   See Mary Griffiths and Simon Dyer, *Upgrader Alley: Oil Sands Fever Strikes Edmonton*, a report published by the Pembina Institute, June 16, 2008.

111   For information on pipeline routes to the US, see National Energy Board (NEB) website, www.neb.gc.ca.

112   NEB website.

113   See British Petroleum's website: http://www.bp.com/sectiongenericartcile.do?categoryld=9005028&contentld=7009098

114   Environment Integrity Project, "Tar Sands: Feeding U.S. Refinery Expansions with Dirty Oil," released June 4, 2008, in collaboration with Environmental Defence Canada. See http://www.environmentalintegrity.org and http://www.envitonmentaldefence.ca

115   Lawrence Martin, "Can Harper Make Canadians Feel Good? The Conservatives are Counting on it," *Globe and Mail*, July 16, 2007.

116 Michael Byers, *Intent for a Nation: A Relentlessly Optimistic Manifesto for Canada's Role in the World* (Vancouver: Douglas & McIntyre, 2007).
117 Quote from then Foreign Affairs minister Maxime Bernier, Government of Canada press statement, Nation Talk, May 14, 2008, http://www.nationtalk.ca/modules/news/article.php?storyid=9520.
118 See quote by Fisheries minister, Loyola Hearn, Nation Talk, May 14, 2008.
119 Cited in John Dillon and Ian Thompson, *Pumped Up: How Canada Subsidizes Fossil Fuels at the Expense of Green Alternatives,* KAIROS publication, 2008, p. 17. Original figures provided by the Pembina Institute.
120 Dillon and Thompson, *Pumped Up*, pp. 17–18.
121 These calculations are based on correspondence between federal finance minister Jim Flaherty and the executive director of the Canadian Churches program KAIROS, Mary Corkery, dated March 26, 2008. In a letter to Ms. Corkery, the minister of finance stated that the annual subsidy to the tar sands industry is approximately 300 million dollars. Ecojustice, an environmental legal action group, has filed access to information requests in an effort to get more detail about this number.
122 A windfall-profits tax is only one of the mechanisms used. Norway, for example, uses a complex mix of measures including direct ownership through its state-owned company, Statoil, the largest player in Norway's oil industry, to achieve the same goal.
123 For oil companies initiating projects after March 19, 2007, the phase out of tax writeoffs under the Accelerated Capital Cost Allowance will be gradual: 90% in 2011; 80% in 2012; 60% in 2013; 30% in 2014; and 25% in 2015 and thereafter.
124 The 2008 OECD Report on Canada calls for the elimination of the remaining tax subsidies to the oil sector, including the 100 percent writeoff of intangible costs for tar sands mine developments and special treatment for exploration and development expenses. The OECD Report also strongly criticizes the option of deducting provincial royalties from the oil industry's federal corporate income taxes, arguing that this undermines the ability of the rest of Canada to benefit from tar sands developments. See Report at: http://www.oecd.org/doc ument/3/0,3343,fr_2649_201185_40732867_1_1_1_1,00.html
125 Jim Stanford, "Building a Diversified, Value Added, Productive Economy," Submission of the Canadian Auto Workers to the Competition Policy Review Panel, Government of Canada, January 2008.
126 Jim Stanford, "Building a Diversified Economy," p. 4.
127 Jim Stanford, "Building a Diversified Economy," p. 9.
128 Jim Stanford, "Building a Diversified Economy," p. 6.
129 Jim Stanford, "Building a Diversified Economy," pp. 7–8.
130 Jim Stanford, "Building a Diversified Economy," p. 10.
131 Jim Stanford, "Building a Diversified Economy," p. 10.
132 According to data provided by energy specialist John Dillon, the oil industry's recent profits are unprecedented. Between 2002 and 2007, the profits of Canadian-based oil companies rose from 14.7 billion to 26.4 billion CD. During the same period, the profits of the big five transnational oil companies rose from 30 billion to 123 billion USD.
133 See chapter 4 of the 2008 OECD Report on Canada's economy at: http://www.oecd.org/document/3/0,3343,fr_2649_201185_40732867_1_1_1_1,00.html
134 For a summary and commentary on the OECD Report by Andrew Jackson in the Progressive Economic Forum, see http://www.progressive-economics.ca/2008/06/

22/the-oecd-and-the-tar-sands/

135 Research on the Harper Team and the revolving door between industry and government prepared for the author by Richard Girard, Polaris Institute.

136 According to Elections Canada, EnCana's total contributions to the new Conservative Party in 2003 (including money contributed to the Canadian Alliance Party) amounted to 107,500 dollars.

137 Stephen Harper, "An Open letter to Ralph Klein," *National Post*, January 24, 2001. It was co-signed by Tom Flanagan, professor of political science and former director of research for the Reform Party; Ted Morton, professor of political science and Alberta Senator-elect; Rainer Knopff, professor of political science; Andrew Crooks, chairman of the Canadian Taxpayers Association; and Ken Boessenkool, policy adviser to Stockwell Day when he was Treasurer of the Alberta Government.

138 For an insight into the close connection between Harper and Flanagan as intellectual and philosophical soulmates, see Marci MacDonald's "The Man Behind Harper," *Walrus*, October 2004. MacDonald traces Harper and Flanagan's intellectual passions back to the economic theories of Friedrich Hayek who heralded free markets as the cornerstone of free societies without government intervention or regulation. She shows how Harper ran the National Citizens Coalition on this philosophy under the motto "More freedom through less government." The article also exposes some of the racist underpinnings of Flanagan's work in his book *First Nations? Second Thoughts* where he argues that "'aboriginals were simply conquered peoples who'd been bested by Europeans with a higher degree of 'civilization'"—and then raises the question of whether Harper shares his views.

139 Tom Flanagan, *Harper's Team: Behind the Scenes in the Conservatives' Rise to Power* (Montreal: McGill-Queen's University Press, 2007), p. 21.

140 Flanagan, *Harper's Team*, p. 23.

141 Jack Aubry and Andrew Mayeda, "The ties that bind; Lobbyists maintain close links to major parties despite tougher regulations," *Ottawa Citizen*, January 22, 2008.

142 See Kristen Anderson's bio at www.globalpublicaffairs.ca

143 Cited in Hester, "Canada as the 'Emerging Energy Superpower,'" p. 10.

144 Hester, "Canada as the 'Emerging Energy Superpower,'" p. 10.

145 Hester, "Canada as the 'Emerging Energy Superpower,'" pp. 7, 9–11.

146 See Jennifer Welsh and Annette Hester, "Superpower?" *Globe and Mail*, February 2, 2008.

147 Brian McKenna and David Ebner, "The Kinder, Gentler Energy Superpower," *Globe and Mail*, January 28, 2008.

148 Cited in Hester, "Canada as the 'Emerging Energy Superpower.'"

149 Stephen Harper, comments at a news conference in New Orleans, April 23, 2008.

Chapter 4: Fuelling America

150 U.S. Geological Survey, USGS World Petroleum Assessment 2000. Accessed electronically at greenwood.er.usgs.gov/energy/World_Energy/DDS-60/index.html. Originally cited in Michael Klare, *Blood and Oil: How America's Thirst for Petrol is Killing Us* (New York, Penguin Books, 2004), p. 16.

151 Klare, *Blood and Oil*, p. 16. Of the 142 billion remaining barrels of oil, 76 billion are "anticipated" additions to known fields and 66 billion are "anticipated" new discoveries.

152 John Blair, *The Control of Oil* (New York: Vintage Books, 1976). See pages 275 ff. Also, see John Dillon, *Power to Choose* (Toronto: GATTFly Publication, 1976),

pp. 12–14. Dillon shows the common interests between OPEC and the big oil transnationals.

153 According to Michael Klare, *Blood and Oil*, British documents declassified in 2004 show that in December 1973, Secretary of Defense James Schlessinger informed the British Ambassador to Washington that the US was considering the use of military operations to seize Saudi Arabia's oil fields if the embargo continued much longer. See Glenn Frankel, "U.S. Mulled Seizing Oil Fields in 73," *Washington Post*, January 1, 2004; and Lizetta Alvarez, "Britain Says U.S. planned to Seize Oil in '73 Crisis," *New York Times*, January 2, 2004.

154 The 1980 rebellion against the Shah of Iran is reminiscent of how ill conceived US oil strategies in the Middle East can backfire in terms of foreign policies, as we see in Iraq today.

155 Hugh McCullum, *Fuelling Fortress America*, p.14.

156 Cited in Klare, *Blood and Oil*, pp. 58–59.

157 Cited Klare, *Blood and Oil*, p. 16. DoE/EIA, AEO 2004, table A11, p. 150.

158 Cited in Klare, *Blood and Oil*, p. 17. Department of Energy's Annual Energy Outlook: DoE/EIA. AEO 2004, table A7, p. 144. DoE/EIA, IEO 2003, table A4, p. 185.

159 Paul Roberts, *The End of Oil* (Boston: Houghton Mifflin Co.; 2004) p. 95.

160 Richard Heinberg, *Peak Everything: Waking Up to the Century of Declines* (Gabriola Island, BC; New Society Publishers, 2007), pp. 135–137.

161 See "The Politics of Oil Security" in Michael Klare, *Resource Wars: The New Landscape of Global Conflict* (New York: Henry Holt & Co., 2001), pp. 29–35.

162 Klare, *Blood and Oil*, p. 11.

163 See Robert E. Ebel, *The Geopolitics of Energy into the 21st Century* (Washington, DC: Center of Strategic and International Studies; 2002), Vols. 1 and 3.

164 Klare, *Blood and Oil*, pp. 28–29.

165 Klare, *Blood and Oil*, p. 33.

166 Klare, *Blood and Oil*, pp. 3–4.

167 Klare, *Blood and Oil*, p. 5.

168 Klare, *Blood and Oil*, pp. 53–54.

169 National Energy Policy Development Group (NEPDG), *National Energy Policy* (Washington, DC: White House, May 17, 2001). Cited in Klare, *Blood and Oil*, p. 57.

170 Klare, *Blood and Oil*, pp. 57–64.

171 US Department of Energy, Energy Information Administration, *International Energy Outlook 2004* (Washington DC, 2004). Cited in Klare, *Blood and Oil*, p. 23.

172 See Michael Klare, "Bush–Cheney Energy Strategy: Procuring the Rest of the World's Oil," *Petropolitics: Oil and Politics* (www.petropolitics. org).

173 Cited in McCullum, *Fuelling Fortress America*, p. 40.

174 Robert Keohane, *After Hegemony: Cooperation and Discord in World Political Hegemony* (Princeton, N.J., Princeton University Press, 2005).

175 See Tony Clarke, "Globo-Petro-Cops," *Resurgence Magazine*, October 2007.

176 Michael Renner, "The New Geopolitics of Oil," *Development*, 2006, vol. 49, pp. 56–63, doi:10. 1057/palgrave.development. 1100273.

177 David Sanger, "China's Rising Appetite for Oil is High on Agenda for US," *New York Times*, April 19, 2006.

178 William F. Engdahl, "Revolution, Geopolitics and Pipelines," *Asian Times Online*, June 30, 2005. See also Renner, "The New Geopolitics of Oil."

179 John Foster, "Afghanistan: A pipeline through a troubled land," a report prepared for the Canadian Centre for Policy Alternatives, June 22, 2008.

180 See Michael Klare, "Containing China," TomDispatch.com, April 18, 2006. See

also Sudha Ramachandran, "China's Pearl in Pakistan Waters," *Asia Times Online*, March 4, 2005.

181   Renner, "The New Geopolitics of Oil."

182   Craig Smith, "U.S. Training North Africans to Uproot Terrorists," *New York Times*, May 11, 2004.

183   It should be noted that Ecuador has recently refused to renew the lease for the US's Manta airbase.

184   See Renner, "The New Geopolitics of Oil," and Sanger, "China's Rising Appetite for Oil."

185   *New York Times Magazine*, March 28, 1999.

186   *New York Times Magazine*, January 5, 2003.

187   McCullum, *Fuelling Fortress America*, p. 14.

188   Cited in John Dillon and Ian Thomson, *Pumped Up*, a KAIROS Research Paper, 2008.

189   Dillon and Thomson, *Pumped Up*, p. 23.

190   Dillon and Thomson, *Pumped Up*, p. 23.

191   *Globe and Mail*, June 14, 2008. By comparison, the cost of producing crude oil from conventional sources ranges between 16.82 and 34.43 dollars. At today's oil prices, the industry's profit margins stand to be substantial if not astronomical.

192   John Dillon, "How NAFTA Limits Our Energy Options" (unpublished paper, 2007).

193   Gordon Laxer and John Dillon, *Over A Barrel: Exiting from NAFTA's Proportionality Clause* (Ottawa: Canadian Centre for Policy Alternatives and the Parkland Institute; 2008), pp. 26–37.

194   Laxer and Dillon, *Over a Barrel*, p. 29.

195   Laxer and Dillon, *Over a Barrel*, pp. 31–32.

196   Laxer and Dillon, *Over a Barrel*, p. 36.

197   Laxer and Dillon, *Over a Barrel*, pp. 36 ff

198   For a discussion of the ExxonMobil case, see Laxer and Dillon, *Over a Barrel*, pp. 38–39.

199   In a note to the author, John Dillon points out that there have been no reports issued by this North American Energy Freedom Commission so far and it appears the Bush administration has not appointed members to it. Presumably these were add-on measures to secure passage of the bill in Congress. In any case, it is safe to assume that the objectives of a continental energy plan are being implemented by Washington through cabinet and executive order.

200   Cited in Marsden, *Stupid to the Last Drop*, pp. 70–71.

201   The year the Energy Policy Act became law, US oil consumption rose by 4.3 percent, the biggest annual increase since 1984. According to British petroleum, it was the largest volume increase on record. Marsden, *Stupid to the Last Drop*, p. 69.

202   Marsden, *Stupid to the Last Drop*, p. 71.

203   The 2007 US Energy Security and Independence Act includes a section that says all federal departments, which presumably includes the Pentagon, will not be permitted to buy oil if its extraction generates more greenhouse gases than conventional oil. However, the interpretation of this section is under some dispute.

204   See http://www.enbridge.com/gateway/

205   Cited in McCullum, *Fuelling Fortress America*, p. 17.

206   Cited in Marsden, *Stupid to the Last Drop*, p. 72.

207   Cited in Marsden, *Stupid to the Last Drop*, p. 72.

208   For a discussion of the CCCE plan, see Tony Clarke, "Corporate Canada: Washington's Empire Loyalists," in Ricardo Greenspun and Yasmine Shamsie, eds., *Whose Canada? Continental Integration, Fortress North America and the Corporate*

Agenda (Montreal, Kingston: McGill-Queen's University Press, 2007), pp. 74–104.

209 In its report to the Tri-leaders Summit in Montebello, August 2007, the North American Competitiveness Council (NACC) called for the establishment of a North American Energy Council composed of government and oil industry representatives and the creation of an Energy Outlook on supply and demand trends over a twenty-five-year period. The NACC does not call for conservation measures but emphasizes instead "technological development" and "secure sourcing" as the means to achieving energy security. The NACC also supported the Canada–Mexico Partnership for temporary exchanges of skilled energy workers between the two countries. For an overview of the NACC's report to the Montebello Summit, see Teresa Healy, "North American Competitiveness Council and the SPP," Canadian Labour Congress Research Paper #44, September 2007.

210 Although NAFTA is the more effective international instrument for making Canadian energy exports to the US compulsory, it is not the only one. The International Energy Agency (IEA), which was initially established after the OPEC oil embargo of 1973, was set up to ensure that the US would receive oil imports in case of another major oil embargo by OPEC member countries. Under the IEA's Emergency Sharing System, if a member country's oil supplies are cut by seven percent or more of their average consumption levels, then other member states are required to share their oil supplies by either reducing their own demand or drawing down their emergency reserves. Being a net energy exporter, Canada would be required as a member of the IEA to increase its oil exports to the US in the case of such an embargo. While the US, as a member of the IEA as well, would be expected "in theory" to allocate its domestic oil to other member countries facing an oil embargo, a US State Department official admitted after the signing of the IEA that, "'in practice,'… only under the most extreme emergency situation would the US ever be called upon to share any of its domestic production with other IEA countries." Ed Shaffer, *Canada's Oil and the American Empire* (Edmonton: Hurtig Publishers, 1983), cited in Dillon, "How NAFTA Limits Our Energy Options."

Chapter 5: Ecological Nightmare

211 Cited in William Marsden, *Stupid to the Last Drop* (Toronto: Alfred A. Knopf, Canada; 2007), pp. 125–26.

212 Marsden, *Stupid to the Last Drop*, p. 126.

213 Dan Woynillowicz, Chris Severson-Baker and Marlo Raynolds, Oil Sands Fever: the Environmental Implications of Canada's Oil Rush, Pembina Institute, November 2005, p.15.

214 Andrew Nikiforuk, "Highway to Hell," *One Earth*, Fall 2007, p. 40.

215 Pembina's forecasts are included in Matthew Bramley, Derek Neabel and Dan Woynillowicz, The Climate Implications of Canada's Oil Sands Development, Pembina Institute Backgrounder, November 2005, p.5. It is worth noting that Pembina's assessment of the carbon output of tar sands oil per barrel may be on the conservative side. The National Energy Board, for example, contends that each barrel of tar sands crude produces 125 kilograms of carbon dioxide, a good deal higher than Pembina's 85.5. The reason for this discrepancy may lie in a difference in assumptions about what kind of fuel is being used.

216 Bengt Soderbergh, Fredrik Robelius, and Kjell Aleklett, "A Crash Programme for the Canadian Oil Sands Industry," *Energy Policy*, volume 35, 2007, pp. 1941–42.

217 Matthew Bramley et al., Climate Implications of Canada's Oil Sands Development.

218 Cited in Hugh McCullum, *Fuelling Fortress America*, p. 34.

219 Jeffrey Simpson, "Alberta remains Canada's square peg on emissions," *Globe and Mail*, March 22, 2008.

220 The Pembina Institute maintains that the energy costs could be lower than the IPCC's estimate.

221 A.E. Farrel and A.R. Brandt, "Risks of the Oil Transition," Berkeley, Institute of Physics Publishing Environmental Research Letters, October 2006.

222 Cited in Hugh McCullum, *Fuelling Fortress America*, p. 34.

223 *Globe and Mail*, January 26, 2008, p. F10.

224 One of Obama's senior advisors, Jason Grumet, was quoted as saying: "If it turns out that the only way to produce those resources [i.e., tar sands crude oil] would be at a significant penalty to climate change, then we don't believe that those resources are going to be part of the long term." Cited in *Maclean's Magazine*, July 21, 2008, p. 28.

225 Dan Woynillowicz, *Troubled Waters, Troubling Trends*, Pembina Institute, May 2006.

226 David Schindler, W.F. Donahue and John P. Thompson, *Running out of Steam? Oil Sands Development and Water use in the Athabasca River Watershed: Science and Market-based Solutions.* University of Toronto Munk Centre for Internationalal Studies and the University of Alberta Environmental Research and Studies Centre, May 2007. p. 1.

227 Woynillowicz, *Troubled Waters, Troubling Trends*, Summary, p. 3.

228 Woynillowicz, *Troubled Waters, Troubling Trends*, p. 4.

229 Golder Associates Ltd, "A Compilation of Information and Data on Water Supply and Demand in the Lower Athabasca River Reach," prepared for the CEMA Surface Water Working Group, 2005, see Table 13.

230 Woynillowicz, *Troubled Waters, Troubling Trends*, Summary, p.6.

231 Bruce Peachy, "Strategic Needs for Energy Related Water Use technologies: Water and the EnergyINet," New Paradigm Engineering, February 2005, 34–36.

232 Interview by Andrew Nikiforuk of Dr. David Schindler, via email, July 27, 2007.

233 Cited in Andrew Nikiforuk, "Liquid Asset," *Report on Business Magazine, Globe and Mail*, March 28, 2008.

234 See *Upgrader Alley: Oil Sands Fever Strikes Edmonton*, A Pembina Institute Report, June 2008.

235 Study conducted in 2007 by the engineering firm Morrison Hershfield for the counties of Strathcona and Sturgeon. Cited in Andrew Nikiforuk, "Liquid Asset."

236 Woynillowicz, *Troubled Waters, Troubling Trends*, Summary.

237 Woynillowicz, *Troubled Waters, Troubling Trends*, p. 4.

238 Cited in Andrew Nikiforuk, "Liquid Asset."

239 Environmental Defence, *Canada's Toxic Tar Sands*, February 2008, p. 3.

240 Environmental Defence, *Canada's Toxic Tar Sands*, p. 3.

241 Kevin Timoney, "A Study of Water and Sediment Quality as related to Public Heath Issues, Fort Chipewyan, Alberta." Prepared for the Nunee Health Board Society, 2007.

242 Mackenzie River Basin Board, "State of the Aquatic Ecosystem Report," 2003.

243 Kevin Timoney, "A Study of Water and Sediment Quality as Related to Public Heath Issues, Fort Chipewyan, Alberta, p. 61.

244 *Fort McMurray Today*, November 20, 2007.

245 Andrew Nikiforuk has a graphic description of these mining operations and the equipment used in his "Canada's Highway to Hell," *One Earth*, fall edition, 2007.

246 Rainforest Action Network and ForestEthics, "Bankrupting the Future," March

2006.

247 Mark Anielski and Sara Watson, "The Real Wealth of the Mackenzie Region: Assessing the Natural Capital Values of a Northern Ecosystem." The Canadian Boreal Initiative, 2005.

248 Anielski and Watson, "The Real Wealth of the Mackenzie Region."

249 Anielski and Watson, "The Real Wealth of the Mackenzie Region."

250 Canada Peatlands Sensitivity Map, Natural Resources Canada, 2007. The map provides ratings of peatland sensitivity to climate change.

251 Glenn, A.J., Flanagan, L.B., Syed, K.H., and Carlson, P.J. 2006. "Comparison of Net Ecosystem CO_2 Exchange in Two Peatlands in Western Canada with Contrasting Dominant Vegetation, Sphagnum and Carex, " *Agriculture and Forest Meteorology* 140 (1–4), pp. 115–135.

252 Alberta Wilderness Association, "McClelland Lake Wetland Complex — Jewel of the Boreal."

253 See Rainforest Action Network and ForestEthics, "Bankrupting the Future," Corporate responsibility Report, 2006.

254 Peggy Holroyd, Simon Dyer, Dan Woynillowicz, "Haste Makes Waste: The Need for a New Oils Sands Tenure Regime," Oil Sands Issue Paper No.4, Pembina Institute, 2007, Introduction, p. 6.

255 Holroyd, Dyer and Woynillowicz, "Haste Makes Waste," p. 6.

256 Holroyd, Dyer and Woynillowicz, "Haste Makes Waste," pp. 17–18.

257 Holroyd, Dyer and Woynillowicz, "Haste Makes Waste," pp. 19–20.

258 Several groups have dropped out of the CEMA multi-stakeholder process because of these problems, notably, the Athabasca Chipewyan and the Mikisew Cree First Nations.

259 Cited in "McClelland Lake Wetland Complex — Jewel of the Boreal," p. 20.

260 McCullum, *Fuelling Fortress America*, p. 33.

261 Cited in Andrew Nikiforuk, "Liquid Asset." At a cost of 25,000 dollars per hectare, the cost of reclaiming 96,000 hectares of mined wetlands in the tar sands region could be as much as 24 billion dollars, depending on the replacement criteria and standards.

Chapter 6: Social Upheaval

262 This description, which coincides with my own observations of Fort McMurray, was cited in Andrew Nikiforuk, "Highway to Hell," *OnEarth Magazine,* Fall, 2007.

263 McCullum, *Fuelling Fortress America,* a report published by the Canadian Centre for Policy Alternatives, the Parkland Institute, and the Polaris Institute, 2006, p. 47.

264 McCullum, *Fuelling Fortress America*, pp. 48–49.

265 McCullum, *Fuelling Fortress America*, p. 49.

266 Cited in Andrew Nikiforuk, "Highway to Hell," *OnEarth Magazine*, Fall, 2007.

267 Environics Research Group, Focus Alberta Survey, March 2007.

268 Diana Gibson, *The Spoils of the Boom: Incomes, Profits and Poverty in Alberta*, Parkland Institute, June 2007, Executive Summary.

269 Gibson, *The Spoils of the Boom.*

270 Gibson, *The Spoils of the Boom.*

271 Gibson, *The Spoils of the Boom.*

272 Cited in McCullum, *Fuelling Fortress America*, p. 50.

273 Diana Gibson, *Selling Albertans Short: Alberta's Royalty Review Panel Fails the Public Interest*, Parkland Institute, October 2007.

274 Media Release, "Premier Stelmach Rejects Expert Advice, Short-changing Alber-

tans by Billions," Pembina Institute, October 26, 2007.

275 See pages 124–125 of the 2008 OECD *Economic Survey of Canada* at www.oecd.org/document/3/0,3343,fr_2649_201185_40732867_1_1_1_1,00.html. It appears that the OECD still maintains that Alberta's new royalty regime, although an improvement on the former regime, still leaves "significant un-captured economic rents." (Andrew Jackson). The OECD also states: "It will be necessary to regularly review the [new] royalty regime … one possibility would be to have a formula whereby parameters are reset in line with competitor country royalty rate changes," p. 125.

276 McCullum, *Fuelling Fortress America*, p. 51.

277 McCullum, *Fuelling Fortress America*, p. 51.

278 McCullum, *Fuelling Fortress America*, p. 52.

279 "Alberta labour group says foreign workers forced to pay illegal fees to get jobs," Canadian Press, November 30, 2007.

280 See Mary Griffiths and Simon Dyer, *Upgrader Alley: Oil Sands Fever Strikes Edmonton*, Pembina Institute Report, June 2008.

281 Lori Theresa Waller, "Passing Out in Upgrader Alley," *The Dominion*, October, 2007.

282 gov.ab.ca/home/documents/investing in our future section5.pdf or http://www.alberta.ca/home/395.cfm

283 Information here cited in William Marsden, *Stupid to the Last Drop* (Toronto: Alfred A. Knopf, 2007), p. 184–194.

284 Marsden, *Stupid to the Last Drop*, p. 185–186.

285 Marsden, *Stupid to the Last Drop*, p. 191.

286 Marsden, *Stupid to the Last Drop*, p. 192.

287 Dan Woynillowicz, Chris Severson-Baker, Marlo Raynolds, *Oil Sands Fever: The Environmental Implications of Canada's Oil Rush*, Pembina Institute, November 2005, p. 15.

288 Richard Heinberg, *The Party's Over: Oil, War and the Fate of Industrial Societies* (New Society Publishers, Second Edition, 2005), p. 128. Heinberg quotes from Walter Younquist, *Geodestines*, p. 216.

289 Extensive work has been done on EROI by Cutler Cleveland at the Center for Energy and Environmental Studies at Boston University and Charles Hall at New York State University.

290 Interview with David Hughes, http://global publicmedia.com/transcripts/827, p. 1.

291 Hughes Interview, p. 2.

292 Hughes Interview, p. 2.

293 Marsden, *Stupid to the Last Drop*, p. 120.

294 Marsden, *Stupid to the Last Drop*, p. 120.

295 McCullum, *Fuelling Fortress America*, p. 29. Nellie Cournoyea, Inuvaluit CEO of the Inuvaluit Regional Corporations, contends there are 64 trillion cubic feet of natural gas in the Mackenzie Delta region of the Beaufort Sea while industry analysts' estimates range between 24 and 36 trillion cubic feet.

296 McCullum, *Fuelling Fortress America*, p. 26.

297 CanWest News Service, December 17, 2007.

298 McCullum, *Fuelling Fortress America*, p. 27.

299 Marita Moll, "Power Speaks to Power under the Nuclear Revival Tent," in Cy Gonick, ed., *Energy Security and Climate Change* (Fernwood Publishers, 2007), p. 64–65.

300 Jamie Hall, *Edmonton Journal*, August 28, 2007.

301 David Ebner, *Globe and Mail*, November 21, 2007.

302 Gordon Laxer, *Freezing in the Dark: Why Canada Needs Strategic Petroleum Reserves*, a report published jointly by the Parkland Institute and the Polaris Institute, January 2007, p. 5.

303 Laxer, *Freezing in the Dark*, p.20–21

304 Laxer, *Freezing in the Dark*, p. 6.

305 Cited in Nick Turse, "The Military–Petroleum Complex," *Foreign Policy In Focus*, Washington DC, March 24, 2008.

306 Naomi Klein, "Baghdad Burns, Calgary Booms," *Nation*, May 31, 2007.

307 See Ricardo Acuna, "Time for Us to Say 'No More Oil for War' to US," *Vue Weekly*, January 17, 2008.

308 Turse, "The Military–Petroleum Complex." See also Nick Turse, *The Complex: How the Military Invades Our Everyday Lives* (Metropolitan Books, Henry Holt & Co. 2008). Note: the original figures in this article are in US gallons which have been converted here to barrels of oil — 1 barrel of oil = 42 US gallons.

309 Turse, "The Military–Petroleum Complex."

310 Turse, "The Military–Petroleum Complex."

311 Research compiled for the author by Richard Girard, Polaris Institute.

312 Sheila McNulty, "Push to Bar Oil Sands to Military," *Financial Times*, March 19, 2008.

313 McNulty, "Push to Bar Oil Sands to Military."

314 See Turse, "The Military–Petroleum Complex."

315 See Michael Klare, *Resource Wars* (New York, Metropolitan Books/Henry Holt & Co., 2001), p. 29.

316 See, for example, Bill Weinberg, "Flashpoint in the Flathead: US–Canada War Looms Over Energy, Water," *World War Report*, November 1, 2007.

317 Cited in Weinberg, "Flashpoint in the Flatland," p. 6.

318 The *Washington Post* provided a chilling reminder of this scenario when it revealed on December 30, 2005, a report called the "Joint Army and Navy Basic War Plan — Red." Stamped with the "secret" seal across its cover, the report outlined plans for US Naval forces to capture the port of Halifax, blockade the Vancouver port and secure control over the Great Lakes. At the same time, the US Army would launch a land invasion to seize control of power plants near Niagara Falls, the strategic nickel mines of Ontario and the railroad centre of Winnipeg. Manoeuvres for seizing control of Montreal, Toronto and Ottawa were also outlined. While some may feel this scenario is a bit too extreme, it is certainly plausible if US vital interests such as oil security are threatened. Indeed, a serious threat to US interests in the tar sands could trigger a military response.

Chapter 7: Resistance Movement

319 See Pembina's website at www.oilsandswatch.org.

320 See Parkland's website at www.ualberta.ca/~parkland/index.html.

321 See Sierra Club's website at www.sierraclub.ca.

322 Greenpeace's website on the tar sands can be found at www.greenpeace.org/canada/en/campaigns/tarsands.

323 See the Indigenous Environment Network's website at www.ienearth.org/energy.html.

324 For information on the Dehcho, see their website at www.dehchofirstnations.com/home.htm.

325 The Canadian Boreal Initiative's website is at www.borealcanada.ca/index-e.php.

326 In a note to the author, Hugh McCullum reports that the Dehcho have approved the expanded reserve and will operate the park on a 50–50 basis with Parks

Canada. See Chapter 8 for more details.

327 For information on the Alberta Federation of Labour activities, see www.afl.org/campaigns-issues.

328 See the CEP website section "Energy Security," www.cep.ca/campaigns/energy_security/energy_security_e.html.

329 Public Interest Alberta's website is at www.pialberta.org.

330 The observations in this and the paragraphs that follow are largely based on conversations with Jessica Kalman, tar sands campaigner at the Polaris Institute, who was born and bred in Alberta.

331 At the 2007 forum, Drs. Schindler, Timoney and O'Connor were present along with representatives of the Indigenous Environment Network, Sierra Club and Pembina and allied groups from outside the province such as Environmental Defence, Toronto, Polaris Institute, Ottawa and Natural Resource Defense Council, Washington.

332 See Polaris Institute's tar sands campaign website at www.tarsandswatch.org.

333 For Canadian Labour Congress 2008 resolutions on climate change and green jobs, see: http://canadianlabour.ca/sites/clc/files/u2/media/15-17sub.pdf and http://canadianlabour.ca/sites/clc/files/u2/media/17-19compost.pdf

334 See KAIROS, Ecumenical Justice Initiatives of the Canadian Churches at www.kairoscanada.org/e/index.asp.

335 Environmental Defence's website is at www.environmentaldefence.ca. See for example, ED's campaign activities on the tar sands at http://www.environmental defence.ca/campaigns/whatsnew/TarSandsAd.htm.

336 See Natural Resources Defense Council's website at www.nrdc.org, and their campaign on dirty fuels and the airline industry at beyondoil.nrdc.org/news/friendlyskies.

337 See http://www.polarisinstitute.org/files/Enviro%20Community %20EISA%20Section%20526%20letter-%20May%207%2008%20-FINAL.pdf.

338 See Indigenous Environment Network's campaign website at: www.ienearth.org/CITSC/Sand_Tar_Campaign.

339 See Rainforest Action Network's website at ran.org/#issues.

340 See Oil Change International's website at www.priceofoil.org.

341 Petr Cizek, "Scouring Scum and Tar from the Bottom of the Pit; Junkies Desperately Seeking One Last Giant Oil Fix in Canada's Boreal Forest," in Cy Gonick, ed., Energy Security and Climate Change: A Canadian Primer (Winnipeg & Halifax: Fernwood Publishers & Canadian Dimension, 2007), pp. 46–68.

342 Cizek, "Scouring Scum and Tar from the Bottom of the Pit," pp. 52–53.

343 For analysis of US foundations and environmental groups cited by Cizek, see Been Brown So Long, It Looked Like Green to Me: The Politics of Nature (Monroe, ME: Common Courage Press, 2004); M. Dowie, American Foundations: An Investigative History (Cambridge, MA; MIT Press, 2002); and F. Pace, Wilderness, Politics, and the Oligarchy: How the Pew Charitable Trust is Smothering the Environment Movement (Counterpunch, 2004).

344 For a sample statement of this "conditional moratorium," see the moratorium declaration posted by the Polaris Institute at www.tarsandswatch.org.

345 Paul Hawken, Blessed Unrest: How the Largest Movement in the World Came into Being and Why No One Saw It Coming (New York: Penguin Group, 2007).

346 Hawken, Blessed Unrest, Introduction.

Chapter 8: Dream Change

[347] Jared Diamond, *Collapse: How Societies Choose to Fail or Succeed* (London: Penguin Books, 2005).

[348] Joseph Tainter, *The Collapse of Complex Societies* (Cambridge: Cambridge University Press, 1988).

[349] Jerry Mander, ed., *Manifesto On Global Economic Transitions: Towards a Global Movement for Systemic Change, Economics of Ecological Sustainability, Equity, Sufficiency and Peace*, A project of the International Forum on Globalization, the Institute for Policy Studies, and the Global Project on Economic Transitions (San Francisco: 2007), pp. 1–4.

[350] Mander, *Manifesto on Global Economic Transitions*, p. 2.

[351] Mander, *Manifesto on Global Economic Transitions*, p. 2.

[352] For a backgrounder on the concept of "sustainability" see Richard Heinberg, *Peak Everything: Waking Up to the Century of Declines* (Gabriola Island: New Society Publishers, 2007), pp. 86–88.

[353] See, for example, William E Rees, Mathis Wackernagel, and Phil Testemale, *Our Ecological Footprint: Reducing Human Impact on the Earth* (Gabriola Island: New Society Publishers; 1995).

[354] Heinberg, *Peak Everything*, pp. 88–95. I have used each of Heinberg's five axioms but have changed their order slightly and made a few minor edits in the wording (in italics) in order to adapt them to our purposes. And I have applied these axioms to the discussion on matters relating to the tar sands.

[355] Heinberg, *Peak Everything*, p. 90.

[356] As the basis for this and several of his other axioms, Heinberg cites the work of Albert A. Bartlett, "Reflections on Sustainability, Population Growth, and the Environment — Revisited," *Renewable Resources Journal*, Vol. 15, No. 4, Winter 1997–1998, pp. 6–23.

[357] This formula was first proposed by petroleum geologist Colin J. Campbell in 1996. For a full treatment of the protocol, see Richard Heinberg, *The Oil Depletion Protocol: A Plan to Avert Oil Wars, Terrorism, and Economic Collapse* (Gabriola Island: New Society Publishers, 2006). See also www.oildepletionprotocol.org.

[358] Heinberg, *Peak Everything*, p. 94.

[359] Heinberg, *Peak Everything*, p. 95.

[360] This ten-point platform is drawn from several sources. The writings of Gordon Laxer of the Parkland Institute have been particularly helpful, especially his "Climate Change and Energy Security for Canadians," in Cy Gonick, ed., *Energy Security and Climate Change* (Winnipeg & Halifax, Fernwood Publishers & Canadian Dimension, 2007) pp. 89–96; and "Freezing in the Dark," Parkland Institute & Polaris Institute, 2008, pp. 27–28.

[361] Lambton Industrial Society, "Deep Well Storage in Salt Caverns — Lambton County," Sarnia, Ontario, revised 1995. www.sarniaenvironment.com/pdf/SLEA-Monograph-L3.pdf.

[362] Quebec and the Atlantic provinces import approximately 850,000 barrels per day, 90 percent of this from OPEC countries such as Algeria, Iraq and Saudi Arabia. Reversing the Sarnia to Montreal pipeline to transport western oil to Eastern Canada could replace 250,000 barrels per day. Add to this the daily oil production of Newfoundland of 368,411 barrels a day (2007), three-quarters of Eastern Canada's oil imports could be replaced by domestic production.

[363] Cited in Laxer, "Climate Change and Energy Security for Canadians," p. 95.

[364] For more details, see Jack Santa-Barbara, "Peak Oil and Alternative Energy," in

Cy Gonick, ed., *Energy Security and Climate Change,* pp. 37–45.

365 See Canadian Centre for Policy Alternatives, *Alternative Federal Budget 2008* (Ottawa: CCPA Publication; 2008), section 4.2, Sectoral Development Strategy, managing the resource boom, pp. 98–100. See also section 2.1, "Climate Change and Carbon Pricing," pp. 67–69.

366 *Alternative Federal Budget 2008,* pp. 67–69.

367 For these and other proposals, see Dale Marshall, "Fudging the Numbers," in Cy Gonick, ed., *Energy Security and Climate Change.* pp. 153–154; also Cy Gonick and Brendan Haley, "A Twelve Step Program to Combat Climate Change," in Gonick, ed., *Energy Security and Climate Change,* pp. 160–165.

368 Jack Santa-Barbara, "Peak Oil and Alternative Energy," in Gonick, ed., *Energy Security and Climate Change,* pp. 44–45. A joule is defined as "the work done, or energy expended, by a force of one newton moving one metre along the direction of the force." It is also defined as the work done to produce the power of one watt continuously for one second. A gigajoule is one billion joules.

369 Santa-Barbara, "Peak Oil and Alternative Energy," p. 44.

370 See Joanna Macy, "The Shift to a Life Sustaining Civilization," on web page "The Great Turning," at www.joannamacy.net/html/great.html. Cited in David C. Korten, *The Great Turning: From Empire to Earth Community* (San Francisco: Berrett-Koehler Publishers & Kumarian Press, 2006), p. 3.

371 Korten, *The Great Turning,* p. 3.

372 Korten, *The Great Turning,* pp. 21–22.

373 Also referred to as Turtle Island, Terra Nulius, Denendeh, Earth Creation, and Peoples' Land.

374 See Dehcho First Nation website at www.dehchofirstnations.com.

375 Information about the Dehcho land-use plan was shared with the author in various meetings and discussions. As well, discussions with Hugh McCullum, who has worked as communications officer with the Dehcho First Nation, were helpful in clarifying the elements of their land-use plan.

376 As noted in Chapter 7, the Dehcho First Nation and Parks Canada are calling on the federal government to establish an expanded Nahanni Park, which the Dehcho and Parks Canada would run on a 50–50 basis. According to a note from Hugh McCullum, "progress is being thwarted by the resource-oriented Government of the Northwest Territories which wants a smaller park which would allow sub-surface resource developers to explore and develop lead, zinc, gold, diamonds, coal, gas and other sub-surface rights. The boundaries of the park will be expanded to protect more than 95 percent of the Greater Nahanni Ecosystem while leaving a small buffer of non-park land around Nahanni Butte and the existing interests of two mining companies — Canadian Zinc and Northern Tungsten — intact until they cease operations in two or three years. In passing the resolution the onus is now on the federal government through Parks Canada to sign an agreement unless its junior colony in the GNWT exerts pressure on behalf of resource developers. At a meeting in Ottawa in April, Grand Chief Gerald Antoine received indications from Environment Minister John Baird that the recommendations will be quickly approved by Ottawa. Twenty-five percent of the highest mineral potential of the Park remains open for development but are restricted to areas where there are existing mineral leases."

377 For some of his more recent writings, see Herman Daly, "Economics in a Full World," *Scientific American,* September 2005; and Herman Daly and Josh Farley, *Ecological Economics: Principles and Applications* (Washington, DC: Island press, 2004).

[378] The following is based on notes taken at a public lecture given by Herman Daly at the Sustainable Energy Forum: Peak Oil and the Environment Conference, Washington DC, May 7–9, 2006.

[379] Jim Stanford, *Economics For Everyone: A Short Guide to the Economics of Capitalism* (Winnipeg: Fernwood Books & CCPA Publishers, 2008), pp. 176–177.

[380] Cy Gonick, "Introduction," *Energy Security and Climate Change*, p. 19.

[381] Based on discussions with Ecuadorians plus an unpublished article by Joan Martinez Alier, "Yasuni in Ecuador: An Initiative from the South."

[382] Alier, "Yasuni in Ecuador," pp. 2–3.

[383] According to Joan Martinez Alier, the formula is as follows: The Ecuadorian government would be asking for compensation of 5 dollars per barrel. 5 x 920 million barrels = 4.6 billion dollars to be put into the public trust fund. At 7.5 percent interest, the 4.6 billion dollar fund would yield 350 million dollars annually.

[384] Similar argument is expressed by Gonick, *Introduction, Energy Security and Climate Change.*

SELECTED BIBLIOGRAPHY

Books

Chastko, Paul. *Developing Alberta's Oil Sands: From Karl Clark to Kyoto*. Calgary: University of Calgary Press, 2004.

Gonick, Cy [ed.]. *Energy Security and Climate Change: A Canadian Primer*. Halifax and Winnipeg: Fernwood Publishing, 2007.

Heinberg, Richard. *The Party's Over: Oil, War and the Fate of Industrial Societies*. Gabriola Island: New Society Publishers, 2005.

Klare, Michael. *Blood and Oil: How America's Thirst for Petrol is Killing Us*. London: Penguin Books, 2004.

Marsden, William. *Stupid to the Last Drop: How Alberta is Bringing Environmental Armageddon to Canada (and Doesn't Seem to Care)*. Toronto: Alfred A. Knopf Canada, 2007.

Monbiot, George. *Heat: How to Stop the Planet from Burning*. Toronto: Doubleday Canada, 2006.

Pratt, Larry. *The Tar Sands: Syncrude and the Politics of Oil*. Edmonton: Hurtig Publishers, 1976.

Roberts, Paul. *The End of Oil: On the Edge of a Perilous New World*. New York: Houghton Mifflin Company, 2005.

Websites

Greenpeace: www.greenpeace.org/canada/en/campaigns/tarsands
Oil Sands Truth: http://oilsandstruth.org/
Parkland Institute: www.ualberta.ca/~parkland/index.html
Pembina Institute: www.oilsandswatch.org
Polaris Institute: www.tarsandswatch.org
Sierra Club: www.tarsandstimeout.ca

INDEX

Abasand Oils Ltd., 28, 29
Aboriginal Pipeline Group Ltd.
   (APG), 200
Abraham, Spencer, 121
Abrams M-1 tank, 210, 213
Accelerated Capital Cost
   Allowance (ACCA), 96, 97
Action Canada Network, 8
Acuña, Ricardo, 219
Adams, Michael, 111
Afghanistan, 132, 207
Africa, 52, 62
Africa, West Central, 132
agro-fuel, 55
Alaska, boreal forest, 174; oil dis-
   coveries, 119; oil revenues, 188
Alberta Agenda, 105–6
Alberta Building Trades Council
   (ABTC), 190
Alberta Chamber of Resources, 167
Alberta Clipper, pipeline, 89
Alberta election 2008, 225
Alberta Energy, 178, 179, 180;
   Energy Efficiency Branch, 150
Alberta Energy and Utilities
   Board, 194
Alberta Environment, 165–66
Alberta Federation of Labour
   (AFL), 101, 190, 191, 222, 226
Alberta government, Commit-
   ment to Sustainable Resource
   and Environmental Manage-
   ment (SREM), 179; land
   reclamation, 181; Regional
   Sustainable Development
   Strategy, 179
Alberta Health and Wellness, 194
Alberta Heritage Fund, 188
Alberta Labour Relations Code, 190
Alberta Oil, 110
Alberta Oil Sands Technology
   and Research Authority, 38
Alberta Progressive Conserva-
   tives, 225
Alberta–US relations, 141
Albian Sands Energy Project, 83;
   tar sands production, 87
Algeria, 204
Alsands consortium, 40, 41
Alternative Federal Budget
   (2008), 266–67
Amazon Basin, 72, 277
Ambrose, Rona, 109
American Association of Petrole-
   um Geologists, 55
American Council on Science
   and Health, 64
American Petroleum Institute, 143
Amoco, 51
An Inconvenient Truth, 59

Anderson, Kristen, 108–9
Andes, glaciers, 62, 277
Anielski, Mar, 137
Antarctica, 52, 60, 63
Arctic, 52, 53
Arctic National Wildlife Refuge
   (ANWR), 122, 277
Arctic Ocean, commercial traffic, 93
Arctic oil, 94
Arctic sovereignty, 93, 215
arsenic, 169, 170, 194
Asia, oil consumption of, 129
Asia–Pacific Economic Co-operation
   (APEC) Summit, 2007, 80, 112
Association for the Study of Peak
   Oil (ASPO), 51, 54
Athabasca Chipewyan, 171, 172, 224
Athabasca Delta, contamination
   of, 170
Athabasca Oil Sands project, 33, 190
Athabasca region, 217, 239, 273;
   land claims, 220
Athabasca River, 16, 77, 162–63;
   165,166, 224, 255; contamina-
   tion of, 170, 193
Athabasca tar sands, 19, 82,
   172–73
Athabasca–Peace–Mackenzie
   watershed, 172
Athapaskan, 20
Atlantic Richfield, 33, 38, 81, 82
Atomic Energy of Canada Ltd., 202
Aurora, 83
Auto Pact, 99
autoimmune diseases, 194

Baird, John, 67, 109, 156
Baku–Tibilisi–Ceyhan pipeline,
   132
Ball, Max W., 28
Bandar, Prince, 120
Bank of Montreal, 226, 243
Baran, Yaroslav, 108
Beaufort Sea, 165; gas reserves, 200
benzene, 170, 194, 195
Berger, Justice Thomas, 34, 232
Bin Laden, Osama, 120
Birch River, 162
bitumen, 16, 17–18; extraction
   of, 22–24, 148; extraction in
   situ, 178, 162; extraction, natu-
   ral gas used, 196–97;
   extraction, strip mining, 163,
   167; upgrading, 86–88, 166
Bitumount, 29–30
Black Creek, 15
Blair, John, 117
Blair, Sidney, 30
Blake, Melissa, 184, 185
Bodman, Samuel, 141, 143–44, 212

Boessenkool, Ken, 105
Bolivia, 62, 96, 133
boreal forest , 173–74
Bradley fighting vehicle, 210
Brandt, Adam, 156
Brazil, 53; water, 73
British North America Act
   (BNA), 26
British Petroleum (BP), 64, 65,
   81, 85–86, 90, 117, 211, 228,
   245; contracts in Iraq, 131;
   defense contracts, 210, 213
Bruce Power, 203
Bruntland Report, 252
Bucyrus, 495HF, 173
Burgan oil field, 53
Burston-Marsteller, 64
Bush administration, 9, 60, 95,
   135, 136, 141, 216, 258; Iraq, 207
Bush Doctrine, 128, 133–134
Bush family, oil connections, 120
Bush, George H., 119, 127
Bush, George W., 55, 112, 115,
   120, 121, 122, 124, 127, 130,
   140, 160, 208, 212
Business Council on National
   Issues (BCNI), 43, 45, 144. See
   also Canadian Council of Chief
   Executives
Byers, Michael, 93

cadmium, 195
Calderón, Felipe, 112
California, low carbon fuel regu-
   lations, 159, 227
Campbell, Colin J., 51, 55–56
Canada Lands Act, 39–40, 95
Canada, Arctic sovereignty, 215;
   oil reserves, 136; oil supplies to
   US, 137; water, 73–74
Canada–United Kingdom Cham-
   ber of Commerce, 80, 92
Canada–US Free Trade Agree-
   ment (FTA), 45–47, 98,137–38,
   144; Article 904, 45–46
Canadian Alliance Party, 105, 150
Canadian Association of Petrole-
   um Producers, 19, 137
Canadian Bar Association, 158
Canadian Boreal Initiative, 176,
   221; corporate donors, 229
Canadian Centre for Energy
   Information, 199
Canadian Centre for Policy Alter-
   natives, 266
Canadian Constitution, 26
Canadian Council of Chief Exec-
   utives (CCCE), 144, 146
Canadian Defence and Foreign
   Affairs Institute, 110

Canadian Development Corpora-
tion (CDC), 36–37
Canadian dollar, 99
Canadian Imperial Bank of
Commerce (CIBC), 56, 226,
243
Canadian Institute of Resources
Law, 180
Canadian Labour Congress, 226
Canadian Military, Afghanistan, 132
Canadian Natural Resources Ltd.
(CNRL), 81, 84, 191, 194; hir-
ing practices, 189–90; tars
sands production, 87
Canadian Oil Sands, 82–83
Canadian Ownership Charge, 40
Canadian Parks and Wilderness
Society (CPAWS), 177, 221–22,
226, 231; Pew Charitable
Trusts, 229
Canadian Petroleum Association
(CPA), 42
cancer, 194
CANDU reactor, 202
CANMET Energy Technology
Centre, 169
carbon capture and sequestration
(CCS), 155–56, 239
carbon emissions, intensity-based
reduction targets, 152
carbon storage, boreal region,
174–76; Mackenzie region, 176
carbon tax, 68, 159, 264, 266
Carlyle Group, 120
Carnegie Institution, 182
Carney, Pat, 45
Carter Doctrine, 126, 127
Carter, Jimmy, 38–39, 118–19, 127
Caspian Sea, 132, 136
Caterpillar 797B, 173
Cato Institute, 64, 136
CD Howe Institute, 216
Center for Strategic and Interna-
tional Studies, 110, 136
Center for the Study of Carbon
Dioxide, 64
Central Asia, 62
Central Command (Centcom), 127
Central North American Water
Project, 76
Chad, 132
Chard, 84
Chavez, Hugo, 136
Cheney Report, 121–22, 123,
127–28
Cheney, Dick, 120, 121–22, 130, 136
Cherry Point Refinery, 211
Chevron, 64, 117, 120, 141–42;
US defense contracts, 213
Chevron Canada, 83, 109; tar
sands production, 87
ChevronTexaco, 83, 121
China, 52, 62, 131, 132, 133; bid for
Unocal, 141–42; competing for tars

sands oil, 214–15; oil consumption,
128–29; relations with Iran, 131;
relations with the US, 129, 131;
water, 73
China National Offshore Oil
Corporation, 141
China National Oil and Gas
Exploration and Development
Corp., 142
Chrétien government, 60, 95,
107, 157; Iraq, 207; Kyoto, 66
Chrétien, Jean, 43, 188
Christian Labour Association of
Canada (CLAC), 190–91
Christina Lake, 83
Citibank, 228, 243
Cities Service, 31, 33, 38, 81
Cizek, Petr, 229
Clark, Joe, 39
Clark, Karl, 21, 22, 29
climate change, 59–61, 149–50,
258; effects of, 62–63
Climate Change Accountability
Act, 243
Clinton, Bill, 119, 127
Clinton, Hillary, 112
coal-bed methane, 201–2, 239
Cold Lake, 40, 85, 211, 273
Cold Lake region, 19, 82, 172–73,
217; land claims, 220
Cold War, 116, 129
Colombia, 133
Columbia University, 49
Communications, Energy and
Paperworkers Union of Cana-
da (CEP), 101, 222, 226
Concerned Citizens Advocating the
Use of Sustainable Energy, 203
Conference on Climate Change,
2007, 174
ConocoPhillips, 81, 82–83,
83–84, 85, 90, 105, 109, 229; tar
sands production, 87; US
defense contracts, 211, 213
ConocoPhillips Canada, 105, 201
Conservative Party of Canada,
donation from EnCana, 105
Consolidated Mining and Smelt-
ing Company, 28
Continental Oil, 33
Council of Canadians, 227
Council of Canadians, Prairies
Region, 223
Cumulative Environmental Man-
agement Association (CEMA),
179–80, 224–25

Daly, Herman E., 275
Daphne Island, 33
Darfur, 133
Darwish, Leila, 219
David Suzuki Foundation, 66, 152
Davis, Bill, 43
Day, Stockwell, 105

Dehcho First Nation, 171, 221,
222, 225; land-use plan,
273–75
Dehcho River, 17
Dene Nation, 8, 17, 20, 27, 34, 232
Diamond, Jared, 249
Diefenbaker government, 24, 92
Diefenbaker, John, 93
Dillon, John, 138
Dion, Stéphane, 66, 67–68,
158–59
Djibouti, 133
Dover, 84
Dow Chemical, 191
Drake, Edwin L., 14
Dyer, Simon, 219

Earnscliffe Strategy Group, 107–8
Earth Summit, Rio de Janeiro, 60
eastern Canada, dependence on
foreign oil, 204
Ebner, David, 111
ecological footprint, 253
Ecuador, 62, 96, 133, 277, 278
Eisenhower, Dwight D., 126, 130, 216
elephant oil fields, 53
Ells, Sidney, 19–21, 22, 28
Enbridge Pipelines of Canada,
83–84, 89, 105, 108, 204, 211,
142, 105
EnCana Corporation, 83–84,
104–5, 229; tar sands produc-
tion, 87
Energy Alberta, 202
Energy Breakthrough Project, 37
Energy Independence and Secu-
rity Act, 160, 227
Energy Resources Group, 156
Energy Security Corporation, 38
Enron, 120–21
Environment Canada, 66, 197
Environment Integrity Project
(EIP), 90
Environmental Defence, 169, 170,
171, 182, 226
Environmental Defense (US), 65
EROI (energy-return-on-invest-
ment), 197–98
European Union, carbon emis-
sions, 155
Exxon, 33, 64, 81, 117
Exxon Valdez, 169
ExxonMobil, 64, 81, 82–83, 84,
121, 200, 201; contracts in Iraq,
131; legal challenge, 140; tar
sands production, 87; US
defense contracts, 210, 213

Farrell, Alex, 156
Field, Chris, 182
Fisheries Act of Canada, 242
Fisheries Canada, 166
Flaherty, Jim, 96
Flanagan, Tom, 106

Foote, Lee, 181
Foreign Investment Review Agency (FIRA), 36–37, 41, 42, 44
foreign workers in the tar sands, 239
ForestEthics, 173, 228
Fort Chipewyan, 26, 193, 219, 221, 224, 242; contamination of wildlife, 195; health concerns, 193–94, 244; health study, 194–95
Fort Chipewyan Community Health Authority, 170
Fort Hills Energy Project, 84, 177; tar sands production, 87
Fort McKay, 193, 221
Fort McMurray, 17, 188, 217, 241; effects of the tar sands boom, 184–86
Fort Saskatchewan, 193
Foster Lake, 83
Friedman, Thomas, 134
Friends of Science, 150
Friends of the Earth International, 228–29

G-8 nations, carbon emission reductions, 154
G-8 Summit of Industrialized Nations (2006), 79
G-8 Summit of Industrialized Nations (2008), 161
Gates, Robert, 212
Gateway Pipeline, 221, 240, 241
Gault, Sebastian, 110
Geological Survey of Canada, 198
George, Rick, 144
Georgia, 132
Ghawar oil field, 53
Gibson, Diana, 186, 219
Global Climate Coalition (GCC), 64, 65
Global Forest Watch Canada, 174
Global Public Affairs, 108–9
Gnome Project, 24
Golder report, 195
Gonick, Cy, 276
Gore, Al, 59
Government of Canada, Mines Branch, 19
Granatstein, J.L., 216
Great Canadian Oil Sands, 32, 33, 37, 83
Great Depression, 28, 29, 66, 279
Great Lakes, 73
Great Recycling and Northern Development Canal plan, 76
Great Slave Lake, 165
Green Party of Canada, 237
Green Shift Plan, 68, 158–59
Greenland, 62, 63
Greenpeace, 219, 220
Guantanamo Bay, 213
Gulf Canada, 38
Gulf Oil, 33, 81, 117

Gwi'chin First Nation, 171

Halliburton Oil Services, 120
Hansen, James, 59, 63
Harper government, 93, 104, 107, 111; Afghanistan, 207; 2007 budget, 97; carbon emission reduction targets, 152–54; Climate Change Technology Fund, 158; energy security, 206; Kyoto, 66–67; Mackenzie Gas Project, 201; Major Projects Management Office, 143; Vision for the North, 94–95
Harper, Stephen, 10, 67, 79–80, 92–93, 103, 110, 112, 142–43, 158–159, 168; connection to the oil industry, 104–8; firewall letter, 105–6
Hawken, Paul, 247
Hayek, Frederich, 106
Health Canada, 194
Heinberg, Richard, 57, 124, 197, 253–57
Heritage Foundation, 64
Hester, Annette, 110–11
Hibernia, 15, 103, 140, 261
High Arctic, natural gas, 199–200
Highway 63, 184
Hill & Knowlton, 105
Himalayas, glaciers, 62
Hirsch report, 53, 56–57
Hirsch, Robert L., 51
Hodson, Dr. Peter, 172
Home Oil, 33
Horizon Project, 190
House of Commons Natural Resource Committee, 172
Houston Summit, 143
Howe, C.D., 28
Hubbert, Marion King, 49–51, 118
Hudema, Mike, 220
Hudson's Bay Company, 27
Hughes, David, 198–199, 202
Husky Energy, 85, 229; tar sands production, 87
Hussein, Saddam, 209
Hutchinson Whampoa, 85
hydrological cycle, 70–71

Ice Age, 63
Ignatieff, Michael, 135
Imperial Oil, 15, 29, 31, 33, 38, 40, 43, 64, 81, 82–83, 84, 109, 200, 223; tar sands production, 87
Imperial Oil Resources Ventures Ltd., 200
in situ, 23, 162
India, 52, 131; oil consumption, 128–29; relations with Iran, 131
Indian Brotherhood of the Northwest Territories, 27
Indigenous Environment Network, 221, 227

Indonesia, 73
International Boreal Conservation Campaign (IBCC), 174, 175
International Conference on Climate Change, Bali (2007), 155, 161
International Energy Agency (IEA), 52, 53–54, 205; World Energy Outlook, 52
International Forum on Globalization, 48
International Monetary Fund, 277
Inuvialuit, 171, 200
Investment Canada, 44, 101, 266
Iran, 117, 131, 132
Iran, oil reserves, 51
Iran, Revolution (1978–79), 126
Iran, Shah of, 119
Iran–Iraq War, 127
Iraq, 57, 111, 117, 131, 135–36, 137, 204; oil reserves, 51; US invasion of (2003), 9, 114, 128, 130, 207, 209, 216
Israel, 34

Jackpine Mine, 83
Joint Review Panel, 201
Joslyn Mine, 85
JP Morgan Chase, 228, 243

KAIROS, 226
Kalman, Jessica, 224
Kananaskis, 232
Ka-shing, Li, 85
Kearl Lake project, 84, 223
Keepers of the Water, 224
Kennedy, John F., 126
Keohane, Robert, 130
Keystone Pipeline, 89, 101, 222, 240, 241
King, Mackenzie, 28
Klare, Michael, 123, 125, 128, 214
Klein government, 96, 185; oil revenues, 188
Klein, Naomi, 209
Klein, Ralph, 105, 143, 150, 157–58, 188, 200, 223
Korten, David, 270
Kravãík, Michal, 70–71
Kunstler, James Howard, 77–78
Kuwait, 53, 117, 118, 127; oil reserves, 52
Kvisle, Hal, 203
Kyoto, 60, 64–65, 66,109, 149, 150, 152,157, 222, 228, 264

Lac Cardinal, 202
Lake Athabasca, 169
Lalonde, Marc, 39, 41, 42–43
Lambton County, 15
land reclamation, 180–181, 239
Latin America, oil-producers, 133
Laxer, Gordon, 138, 206, 219
Layton, Jack, 243
Ledcor Group, 191

Leduc, 15, 29
Lewis, 84
Lexington Project, 124
Liberal Party of Canada, 158–159; Green Shift Plan, 68, 158–59
Lloydminster, 85
Lougheed government, 188
Lougheed, Peter, 34–35, 37, 40, 43, 47, 81–82, 158, 233; tar sands moratorium, 232
Lubicon First Nation, 221
Lunn, Gary, 202

MacDonald, Donald, 45
MacEachen, Allan, 42
Mackay River, 84
Mackenzie Gas Project (MGP), 94, 95, 200–1, 221
Mackenzie River , 17, 162, 165, 171, 221. See also Dehcho River
Mackenzie River Basin Board, 169, 171
Mackenzie Valley Pipeline, 8, 33–34, 277; Royal Commission Inquiry, 232
Mackenzie, Alexander, 17
Macy, Joanna, 270
Mander, Jerry, 48–49, 250
Manhattan Project, 24
Manning government, 29–31
Manning, Ernest, 24, 30
Manning, Preston, 106
maquiladoras, 71–72
Marathon Oil, 83; tar sands production, 87
Marie Lake, 219
Marsden, William, 22
Martin government, 93, 95, 107
Martin, Lawrence, 93
McCain, John, 124
McCarthy, Shawn, 110
McClelland Lake Wetland Complex, 177
McCullum, Hugh, 181, 201–2
McGowan, Gil, 222
McKenna, Brian, 111
McQuaig, Linda, 110
Meadow Creek, 84
mercury, 169, 195
Mexico, 53, 71–72, 262; NAFTA, 138
Middle East, 209
Middle East Crisis (1973), 35
Mikisew Cree, 171, 172, 224
Mikula, Randy, 168–169
Mildred Lake, 28, 33, 83
Mill, J.S., 276
Mobil, 64, 117
Monbiot, George, 60, 61
moratorium, tar sands development, 265; versions of, 233–36
Morgan, Gwyn, 104–105, 109
Morrow, Justice William, 27, 34

Mueller, Clayton Thomas, 221
Mulroney, Brian, 43–44, 45
Muskeg River, 170
Muskeg River Mine, 83

Nahanni National Park, 222
Nahanni River, 222
napthenic acids, 169
NASA Goddard Institute for Space Studies, 59
National Citizens Coalition, 105
National Council on Welfare, 187
National Energy Board (NEB), 42, 46, 89, 201, 202, 222, 206, 266
National Energy Board (US), 199
National Energy Policy Development Group (NEPDG), 120, 121–22
National Energy Policy Development Group Report. See Cheney report
National Energy Program (NEP), 35, 39–40, 41, 42, 43, 47, 48, 95, 104, 157, 158, 159, 204, 259; legacy of, 218
National Environmental Policy Institute, 64
National Geographic, 57
National Oil Policy (1961), 31–32
National Post, 106
National Round Table on the Environment and Economy, 154–55, 155–56, 158
National Wetland Coalition, 64
Natland, Manley, 23–24
natural gas, 199–200; reserves, 198–99, 239
Natural Resource Defense Council, 227
Natural Resources Canada, 66, 169, 198, 200, 202
Nature, 181–82
New Democratic Party of Canada, 237, 243
New Energy for America program, 124, 268
New Royalty Framework, Alberta, 189
New York Times, 142, 214
Nexen Oil Sands, 82–83
Niagara Institute, 43
Nigeria, 57, 111, 132, 136
Niglintgak gas field, 201
9/11 Attacks, 9, 120, 128, 130
Nixon, Richard, 118
Norquay, Geoff, 107–8
North American Energy Working Group, 146
North American Free Trade Agreement (NAFTA), 8, 98, 111, 112–13, 138–40, 144, 146–47, 216, 261, 262; Article 315, 138; Article 605, 138;

Chapter 11, 139–40, 144; proportional sharing clause, 144, 138–39, 206, 227, 262
North American Leaders' Summit, Montebello (2007), 146
North American Leaders' Summit, (2008) 112
North American Security and Prosperity Initiative, 144
North American Water and Power Alliance, 76
North Saskatchewan River, 166–67, 191, 192–93
North Sea oil fields, 53, 119
Northern Lights project, 85
Northwest Passage, 17, 63, 92, 215
Northwest Territories, Supreme Court, 27; resource development, 225
Norway, 57, 264; oil revenues, 188
NRDC, 228
nuclear power, 239; and tar sands development, 202–4

O'Connor, Dr. John, 193–94, 195
Obama, Barack, 112, 124, 160, 216, 268
Oil and Gas Journal, 136
Oil Change International, 228
Oil Depletion Protocol, 254, 261
oil, deep sea, 54; offshore, 52; strategic regions, 131–33
oil-exporting countries, 117
oil gum beds, 15
Oil Sands Conference, 1951, 30
Oil Sands Experts Group, 146
Oil Sands Truth, 226
Oil Shockwave, 57
Oil Springs, 15
Oilweek, 42
Old Man River, 77
Ominayak, Chief Bernard, 221
Operation Southern Watch, 127
Opti-Nexen upgrader, 88
Organization for Economic Co-operation and Development (OECD), 97, 189; survey of Canada's economy, 102
Organization of Petroleum Exporting Countries (OPEC), 34, 51–52, 57, 119, 204; formation of, 117; oil embargo (1973), 34, 35–36, 117–18

PAHs (polycyclic aromatic hydrocarbons), 169, 170, 194
Pakistan, 62
Parkland Institute, 186, 188, 189, 205, 206, 218, 219, 224, 227
Parks Canada, 222
Parsons Lake gas field, 201
Partnership for Climate Action, 65
Peace region, land claims, 220

Peace River, 77, 165, 170
Peace River region, 162, 217, 273;
    nuclear power plant, 202–203;
    tar sands, 19, 82, 172–73
Peace–Athabasca Delta, 163
Peachy, Bruce, 164
peak oil, 10, 49–59, 258, 279
Pearle, Richard, 130
peatland, 174, 175, 176–77, 178
Pembina Institute, 151, 157,
    162–63, 178, 179, 181, 196,
    218, 219, 223, 224, 226; con-
    necting the drops, 224;
    corporate donors, 229; declara-
    tion on the tar sands, 231; tar
    sands moratorium, 232
Pentagon. See US Department of
    Defense,
permafrost, 63, 175, 177, 181
Persian Gulf, 131–32, 127, 141
Peru, 62
Petro-Canada, 36–37, 39, 40, 47,
    81, 82–83, 84, 109, 177, 229; tar
    sands production, 87
PetroChina International Co.,
    142, 215
Petrofina SA, 33
Petroleum and Gas Revenue Tax,
    40
Petroleum Incentive Payments
    program, 40
Pew Center for Climate Change,
    65
Pew Charitable Trusts, 229
Pew, John Howard, 32, 41, 83, 229
Phillips Petroleum, 33
pipelines, 89; Gateway (Edmon-
    ton to Kitimat), 142, 215; maps,
    91, 145; Sarnia, 36; Sarnia to
    Montreal, 204, 261; Venezuela
    to Tumaco, 133;
Polaris Institute, 205, 224, 226,
    227
polycyclic aromatic hydrocar-
    bons. See PAHs
Pond, Peter, 16, 25
Potsdam Institute for Climate
    Impact, 61
Prairie Acid Rain Coalition, 223
Pratt, Larry, 7, 22, 30, 81, 92
Prentice, Jim, 201, 202
Progressive Conservative Party
    Alberta, 150
Project for a New American Cen-
    tury, 130, 209
Project North, 8
Prudhoe Bay, 24, 201
Public Interest Alberta (PIA),
    223, 224

Radke report, 192–93
Radke, Doug, 166
Rainforest Action Network
    (RAN), 173, 228

Rapid Deployment Joint Task
    Force, Persian Gulf, 127
Reagan administration, 44, 127
Reagan, Ronald, 41, 42, 44–45,
    119
Rees, William, 252
Reform Party, 105, 150
Regional Aquatics Monitoring
    Group (RAMP), 171–72
Ricardo, David, 276
Rice, Condoleezza, 120, 212
Richfield Oil, 23–24, 31
Roberts, Paul, 123–24
Rocky Mountains, glaciers, 165
Romm, Joe, 52
Roosevelt, Franklin D., 125
Royal Bank, 226, 243
Royal Commission on Economic
    Union and Development
    Prospects, 45
Royal Proclamation of 1763, 26
Royal Society, 62
royalty regimes, 187–89, 264–65
Royalty Review Panel, Alberta,
    189
Rubin, Jeff, 56
Rumsfeld, Donald, 130, 208, 213
Russia, 4, 96, 111, 132; Arctic sov-
    ereignty, 215; boreal forest, 17;
    relations with US, 131; water,
    73
Ruth Lake, 28

Sable Island, 15, 103
Sahtu First Nation, 171, 200
Santa-Barbara, Jack, 269
Sarnia, 36, 90, 204, 261
Saskatchewan Party, 225
Saskatchewan, resource develop-
    ment, 225; tar sands, 19;
    uranium, 203
Saud, Royal House of, 120, 126
Saudi Arabia, 9, 18, 52, 53, 111,
    114, 117, 118, 127, 131, 135,
    137, 204; oil reserves, 51–52,
    135; US national oil security,
    126
Scandinavia, boreal forest, 174
Schindler, Dr. David, 77, 164–65,
    175, 192, 202–3
Science, 182
Scientific American, 51
Scotia Bank, 226, 243
Security and Prosperity Partner-
    ship (SPP), 111, 144, 146, 147,
    216, 223, 227; creation of, 146
seven sisters, 117
Shah of Iran, 38
shale oil, 53, 54, 55
Shamrock Summit, 44–45
Shell Canada Ltd., 33, 108, 201;
    tar sands production, 87
Shell Oil, 31, 49, 64, 65, 81, 83,
    90–91, 117, 191, 192, 194, 228,

229, 245; contracts in Iraq, 131;
    oil reserves, 52; US defense
    contracts, 210, 213
Simmons & Company Interna-
    tional, 54
Simmons, Matthew, 54, 55, 135
Simpson, Jeffrey, 155
Sinopec, 85, tar sands produc-
    tion, 87
Six Nations of the Iroquois Con-
    federacy, 252
Slave Lake, 170
Slave River, 170
Slave River Delta, 165, 170
Smith, Adam, 276
South China Sea, 132
South Nahanni Watershed
    National Park Conservancy,
    274
South Saskatchewan River, 77
Soviet Union, 129; invasion of
    Afghanistan, 126–27
Stanford, Jim, 49, 97–98, 99, 100,
    276
Steepbank River, 195
Stelmach government, 96, 243;
    carbon emission reduction tar-
    gets, 154
Stelmach, Ed, 159, 168, 189, 220,
    225, 233
Stern Report, 65–66
Stern, Sir Nicholas, 65–66
strategic petroleum reserves, 205
Strathcona, 219
strip mining, 22–23, 172–73
Suave, Dr. Michael, 194
Sudan, 133, 136
Sun newspaper chain, 224
Sun Oil Company, 32, 37, 41, 83,
    229
Suncor, 21, 28–29, 65, 81, 83,
    86–87, 105, 144, 170, 190, 191,
    229; contamination report,
    195; land reclamation,
    180–181; oil spill,193
Sunrise, 85
super giant oil fields, 53
Supreme Court of Canada, on
    regulation of emissions, 159
Surmont Project, 83–84, 85
sustainability, 253–57
Syncrude, 21, 24, 32–33, 35, 37,
    38, 81, 81–82, 84, 86–87, 92,
    142, 158; dealings with China,
    215; duck massacre, 244; land
    reclamation, 180, 181; oil spill,
    193
Syncrude Remission Order, 95
Syncrude Tailings Dam, 167–68
Synenco Energy, 85, 109

Tactix Government Consulting,
    108
Taglu gas field, 200

tailings ponds, 167–68, 194; contents of, 169; duck massacre, 168, 244; leakages, 169, 170–171
Tainter, Joseph, 249
Tar Island, tailings pond leaks, 170; upgrader, 83
tar sands boom, distribution of wealth, 186–187

tar sands, carbon emissions, 151–52; extent of development, 178–79; labour, 189–90; production, 143; regulation of, 143; subsidization of, 157
tax writeoffs, 95–96
Taylor, Amy, 219
TechCentralStation, 64
Technology Canada Partnerships, 99
Teck Cominco, 109, 177; tar sands production, 87
Telfer, Lindsay, 219
Teller, Edward, 24
Temporary Foreign Worker Program, 191
Terra Oils, 140
Texaco, 51, 64, 65, 117
Three Gorges Dam, 168
Timoney, Dr. Kevin, 170, 172
Titusville, Pennsylvania, 14
Tlicho First Nation, 171
Toronto Dominion Bank, 226, 243
Toronto Star, 107
Total E&P Canada, 84
Total SA, 83–84, 85; contracts in Iraq, 131; tar sands production, 87
Toxics Watch Society of Alberta, 223
Trailbreaker pipeline project, 204
Trans Mountain Pipeline, 89
TransCanada Corporation, 210, 203
TransCanada Pipelines Ltd., 89, 200
Trans-Sahara Counterterrorism Initiative, 133
treaties of 1921, 274
Treaty 8, 26–27
Trudeau government, 38–39, 42–43, 45, 204, 232
Trudeau, Pierre, 34–35, 36
Truman, Harry S., 126
Tucker Lake, 84, 85
Turner Valley oil fields, 29

UC Berkeley, 49
UK Meteorological Office, 61–62
United Airlines, 227
United Arab Emirates, oil reserves, 51
United Nations, 59, 71, 125, 278

United Nations, Commission on the Limits of the Continental Shelf, 93; Intergovernmental Panel on Climate Change (IPCC), 59–61, 64, 149, 156; Security Council, 133
United States, Arctic sovereignty, 215; control over oil supply lines, 131; Energy Independence and Security Act (2007), 141, 212; Energy Policy Act (2005), 140-41, 142, 212; foreign oil consumption, 119–20; Lend Lease Act, 126; military bases in Africa, 132–33; military bases South America, 133; military spending, 130; National Energy Policy, 122; oil consumption, 123, 268; oil imports, 124–25, 127–128; oil reserves, 116–17, 118; War on Terror, 130; water, 73–74;
United States Atomic Energy Commission (AEC), 24
United States Commission on North American Energy Freedom, 140
United States Department of Defense, 120, 227; alternative energy, 213; dependence on tar sands oil, 140–41; oil consumption of, 207, 209–10
United States Department of Energy, 51, 123, 128
United States Department of Energy and Natural Resources Canada Summit, 2006 (Houston Summit), 142–143
United States Department of the Interior, 168
United States Energy Information Administration, 208
United States Geological Survey, 49, 94, 116
United States Mayors' Association, 240
United States National Commission on Energy Policy, 57
United States National Security Council, 213–14
United States Secretary of the Navy, 25
United States Senate Energy Committee, 59
University of Alberta, 22
Unocal, 141–42
Upgrader Alley, 88, 191–92, 219, 239, 241
upgrading, bitumen, 86–88, 166
Uranium city, 203
Urban Water Council, 74
UTS Energy, 177
UTS Energy Corporation, production actual and projected, 87

VBS.TV, 223
vedoma, 182
Venezuela, 53, 54, 111, 117, 133, 136; oil reserves, 51; relations with the US, 129; tar sands, 103
Vietnam War, 132
Vision for the North, 94
Voyageur upgrader, 195

Wall, Brad, 203, 225
water, contamination of by factory farms, 72; depletion, 68–73; export of, 75–76; use in industry, 70, 71–72
Water Management Framework for the North Saskatchewan, 166–67
water shortages, Alberta, 77; NWT, 77; Saskatchewan, 77; US, 74–75
Waxman, Henry, 212–13
Wells, Jeff, 174
Western Energy Accord, 46–47
Western Sedimentary Basin, 199, 204
Weyburn, 15
White, Thomas E., 121
Williams, James Miller, 15
Wilson, Michael, 45, 212
Wolfowitz, Paul, 130
Wood Buffalo, Municipality of, 184–85
Wood Refinery, Illinois, 227
World Bank, 65, 66, 125, 277
World Commission on Environment and Development, 252
World Meteorological Organization, 61, 62
World Resources Institute (WRI), 73, 152
World Trade Center, 120
World Trade Organization (WTO), 125
World War I, 21, 205
World War II, 29, 125, 279
World Wildlife Fund (WWF), 157, 221, 226, 231; Pew Charitable Trust, 229
Woynillowicz, Dan, 219

Yasuní National Park, 277, 278
Yom Kippur War, 34, 117

Zoellick, Robert, 121